ROUTLEDGE LIBRARY EDITIONS: CONSERVATION

Volume 4

MAN AND WILDLIFE

MAN AND WILDLIFE

L. HARRISON MATTHEWS

Routledge
Taylor & Francis Group
LONDON AND NEW YORK

First published in 1975 by Croom Helm Ltd

This edition first published in 2020
by Routledge
4 Park Square, Milton Park, Abingdon, Oxon OX14 4RN
605 Third Avenue, New York, NY 10017

Routledge is an imprint of the Taylor & Francis Group, an informa business

British Library Cataloguing in Publication Data
A catalogue record for this book is available from the British Library

ISBN: 978-0-367-43303-1 (Set)
ISBN: 978-1-00-300237-6 (Set) (ebk)
ISBN: 978-0-367-41668-3 (Volume 4) (hbk)
ISBN: 978-0-367-81561-5 (Volume 4) (ebk)

Publisher's Note
The publisher has gone to great lengths to ensure the quality of this reprint but points out that some imperfections in the original copies may be apparent.

Disclaimer
The publisher has made every effort to trace copyright holders and would welcome correspondence from those they have been unable to trace.

Man and Wildlife

L. HARRISON MATTHEWS

CROOM HELM LONDON

Croom Helm Ltd
2-10 St. John's Road London SW11

ISBN 0-85664-012-5

Set by Red Lion Setters, Holborn, London
Printed by Biddles of Guildford

CONTENTS

I INTRODUCTION

'Nature is the work of the Devil' –
William Blake, 1757-1827

In the beginning man was himself part of wildlife. Some 30,000 years ago, however, he began to draw away – slowly and almost imperceptibly at first, but later with gathering speed until he now lives in an artificial environment of his own making. He has not only separated himself from wildlife, but has become its rival, exploiter and destroyer.

Over two million years and more before that time early species of man were evolving from the stock of australopithecine man-like apes – their brains were getting larger, and they had got up from all-fours to walk upright on their hind legs. At the same time their eye-teeth, the canines, became smaller so that they no longer projected as fangs above the level of the other teeth. This enabled man to chew his food with a grinding, rotary, or side-to-side movement of the jaws; the apes, which have large canines, can crush their food but cannot grind it in the human manner. The canine teeth in apes and monkeys function not so much for catching or tearing up food as for fighting or making a display of force. By comparison man found himself naked and physically defenceless.

Mammals that live in groups generally have some kind of social organisation with a hierarchy or peck-order; furthermore, both social and solitary-living mammals drive strangers of their own species away from their living and feeding territory. In doing so they sometimes fight, using the weapons – such as teeth, horns or antlers – with which they are endowed. Fighting, however, seldom results in the death or serious injury of either combatant – it is more a trial of strength in which the weaker either submits to taking a subordinate place in the hierarchy or, if a stranger, leaves the winner in undisturbed possession of his territory. It frequently happens that in a confrontation there is no fight because the contestants merely display their weapons and threaten each other, whereupon the least stout-hearted submits to the stronger. If the loser flees, the winner chases him to the bounds of his territory but not usually beyond. On the whole this kind of aggressiveness enables social order or territory to be preserved without the animals killing each other, and spreads the population as widely as possible throughout the habitat.

Man, having no canine teeth large enough for use as weapons or in threat display, took to making artificial weapons. Some of the present day apes even make and use tools – wild chimpanzees have been seen stripping the leaves from twigs to make tools for extracting white ants from their nests in order to eat them. It is presumed that primitive man's larger brain gave him the ability to understand that he could make natural pebbles more useful by breaking them to form crude choppers with a cutting edge. He doubtless also used sticks as tools and improved their natural shapes for his purposes, though wooden artifacts have not survived to prove the assumption.

In using his tools, too, man's large thumb gave the power-grip of his hand a much more precise control than that of the apes, and in throwing a missile his adoption of an over-arm, above the shoulder action gave him accuracy of aim and greater force of propulsion – when apes throw things they lob them under-arm with little precision. Once the early species of man had thus made up for their deficiencies in natural weapons they were able to threaten each other with violence in maintaining their social position within their groups or warning intruders off their territory. The use of artificial weapons, however crude, gives so great an advantage over unarmed combat that man found he could use them with deadly effect against his own species as well as in killing most forms of wildlife for food.

Whether the invention of artificial tools and weapons led to the development of man's characteristic aggressiveness, or merely allowed an innate aggressiveness to find expression, can only be guessed. Man's practical use of aggression has certainly separated him from wildlife and led to his civilisation, in contrast to the comparatively peaceful lives of the apes in which the members of social groups know their places and keep to them, and neighbouring groups do not fight each other. Perhaps the nature of his diet also contributes: that of the apes is mainly vegetable matter, though wild chimpanzees do sometimes eat meat; but when man found he was able to kill all sorts of wildlife with his artificial weapons he became largely carnivorous until the invention of agriculture – he has been eating meat for so long that at least two species of tapeworm which infect him when he eats raw or undercooked beef or pork are specific parasites on man as their host.

It is generally agreed that the ancestors of man were creatures of the forest, who were driven to inhabiting more open country when increasing aridity of climate reduced forest cover. In the new environment the bipedal stance gave the advantages of speed in movement, wider visible horizon, and the release of the fore-limbs from locomotive duties so that they were available for the manipulation of all kinds of objects, natural and artificial. Man then found himself in the Biblical position of having 'dominion over the fish of the sea, and over the fowl of the air, and over the cattle, and over all the earth, and over every creeping thing that creepeth upon the earth'. He has exercised his dominion on an increasing scale to his own material advantage but to the detriment of wildlife, until at last his depredations have come full circle so that they act to his detriment as well.

The legend of the Garden of Eden and the doctrine of original sin are not without foundation, for when the ancestors of man were still part of wildlife they were amoral, and it was not until man became separated from wildlife that he could evolve a system of morals. Even today all children are born amoral – greedy, selfish and aggressive; they have to be taught not to indulge in lying, stealing and violence, and generally to behave in ways acceptable to society. Although the theologians' notion that children are conceived and born in sin may not be acceptable, it is obvious that children are born innately sinful in the sense that they lack any moral values, which they have to learn from adults. However well man may have acquired such morals since then for regulating his behaviour towards his fellow men, he has certainly not acquired them in his attitude to wildlife, in which greed for immediate gain with no thought for the future has predominated at all times throughout the human race, with the exception of a small minority.

Until the invention of agriculture and the domestication of animals man was entirely dependent for food and clothing upon wildlife, both wild animals and wild plants. In the long ages of prehistory when man was a hunter of animals and gatherer of edible plant material, his artificial tools and weapons, used as extensions of his limbs and muscles, increasingly helped in his struggle for survival as he improved his skill in making them. Although he was becoming something apart from wildlife man was ever, and still is, like all other living organisms, innately driven to seek individual survival – to eat and to reproduce.

The continuity and 'purpose' of life have been subjects of speculation since man began to think of matters beyond his immediate needs, but it is only within the last few years that Science has shown how life continues and, as a corollary, that it has no purpose. It is now widely known, not only among those who have made a special study of biology, that the chromosomes, the minute structures that form the main part of the nucleus in each cell of the body of practically all animals and plants, are what carry the hereditary characters from parents to offspring – they determine that species 'breed true'. After the discovery, about a century ago, of the way a set of chromosomes passes from each parent to the fertilised egg cell, which then divides repeatedly to form a new individual, it became obvious that chromosomes carry some sort of coded information to guide the development of the offspring. Experiment showed that the information is carried in discrete units, each of which, either separately or in combination with others, controls the production of a specific character. The nature of the hypothetical units, the 'genes', remained unknown until the discovery of the structure of deoxyribonucleic acid (DNA) showed how the code is constructed. When we had the comfortable hypothetical genes the phenomena of heredity seemed comprehensible, but the discovery of the double helix of DNA, far from simplifying the subject, appears to make it infinitely more complicated.

As far as is known the external environment cannot normally exert any influence upon the genetic constitution of the chromosomes, and consequently there can be no inheritance of characters acquired by an individual in response to its environment. Yet it is obvious that in many ways plants and animals are adapted to their environments. Darwin proposed a theory of natural selection to explain how the adaptations came about, and the mode of origin of the innumerable species thus adapted. No two animals are quite alike and consequently some individuals may have a slight advantage over their companions in the business of life, and their chances of survival and leaving offspring may be greater; they will survive and the others will be more easily eliminated by the natural hazards of the environment. Darwin knew nothing of chromosomes, and thought that characters acquired during life could be transmitted to the offspring. He imagined that every cell of the body contained minute things he called 'gemmules' which could be altered by the environment, and he further imagined that the gemmules became incorporated in the germ cells so that acquired characters would be transmitted to the offspring. Later research has shown the fallacy of his ideas, but has also shown that should the characters that give some individuals an advantage over others be caused by a difference in the chromosomes, the characters could be transmitted to their offspring to produce a strain more suited to the environment.

Changes in chromosomes can be produced experimentally by drastic methods, as by subjecting animals to the action of X-rays, and such changes are hereditable, though most of them are of such a nature as to handicap the animals, and would be quickly eliminated by natural selection in a wild population. Changes in the chromosomes are known as mutations, and it is believed that small mutations that do not lead to harmful results sometimes occur in nature, providing new hereditable characters which can be acted upon by natural selection. It has therefore been proposed that the evolution of new species takes place by the action of natural selection on small random mutations, the accumulation of which in the course of time differentiates new species from their ancestors. Small hereditable changes thus produce a differential survival rate, and a very slight differential may be enough to ensure the survival and production of offspring by its possessors.

The term 'natural selection' is in some ways unfortunate because it implies some similarity to the artificial selection by plant and animal breeders who consciously select characters to produce an organism of a desired kind. Natural selection is really natural rejection, or the survival of the fittest – a term introduced by Herbert Spencer, not Darwin. There is no purposive direction in natural selection but merely a blind destruction of those inadequate for survival.

The DNA forming the genes is an enormous molecule arranged as a long chain on which the sequence of nucleotide bases is the code. It consists of two strands along which the bases are arranged in the same sequence, but the two strands are coiled round each other head to tail so that the sequences are reversed. If suitable materials are to hand DNA is self-replicating: the strands separate, and while separating each makes an exact facsimile of its former partner so that the amount of DNA is doubled. There are only four bases, but they can be arranged in groups in an enormous number of alternative ways to produce a code which controls the kind of protein or kind of enzyme that can

can be made in the cell. This it does through the mediation of ribonucleic acid acid (RNA) which carries, by the arrangement of its bases, the coded message from the DNA to the place where the protein is being made in the cell, thus 'telling' it what kind of protein to make. How the cell is told what to do with the proteins when it has made them remains to be discovered, as does the way in which the code transmits such things as instinct now called 'innate behaviour patterns' for short.

The code arrangement in the self-replicating DNA molecule is the physical basis of heredity, and is the means by which the characters of one generation reappear in the next. It is not the characters that are transmitted, but a message which tells the cells of the new organism what to do. The whole of living nature is thus a vast library written in the language of DNA; every cell contains a message telling it what proteins to make, just as every printed page in a book contains a message which, when decoded from black marks on the paper, generally means something to the reader – though not to the page itself. The DNA code is also analogous to a gramophone record in which the wiggles in a spiral groove engraved on a silent lump of plastic tell the machine what music or other noises to make.

In theory the message is copied exactly at every cell division, but in practice it is not surprising that among so many million duplications a 'copyist's error' sometimes creeps in. If, during self-replication, a mistake occurs so that the sequence of bases is altered, the coded message is changed and a mutation is produced. It is possible to imagine conditions in which the probability of such a mistake is enhanced by the nature and quantity of bases that may be available in the cell during self-replication. The nature and quantity of bases available could be determined by external conditions. If such a theoretical situation is ever proved to occur in nature the whole matter of the impossibility of inheriting acquired characters will need radical re-examination.

The self replicating nuclear material, formerly imprecisely termed the 'germ-plasm', that passes from one generation to the next is potentially immortal – if, indeed it is living. It makes a continuous stream running through the generations, that in the germ cells of one generation passing direct to those of the next, unchanged except for fortuitous mutations and the mixing of the genes received from the parents. The body that contains it, whether of plant or animal, is only its temporary habitation; it seems strange that DNA has found it necessary to build such an infinite variety of houses to be abandoned as they become dilapidated. We flatter ourselves that we are the principals and that our germ-plasm is merely part of us, but the true position is the reverse; we, like all living things, are merely nests for sheltering, and vehicles for transmitting, what really matters: the double helix of DNA.

The self-replicating molecule is believed to have arisen by chemical reaction, under the influence of the sun's ultra-violet light, from simpler non-replicating molecules at a remote stage in the history of the earth, when the atmosphere lacked free oxygen and consisted mainly of gases such as ammonia and methane. Once the first molecule replicated itself the evolution of living things was inevitable – the whole extent of life with its millions of species, past, present, and perhaps to come, was implicit, as was every subsequent activity of man. According to the present state of our knowledge of biology and cosmology, the origin and evolution of man and wildlife is as fortuitous and undirected as that of the planet on which they find themselves – it is not love that makes the world go round, but DNA, our inexorable master. DNA will, presumably, go on self-replicating and building up organisms to house itself until the sun suddenly turns into a nova and brings all to an end -- an event that may happen at any moment or may be postponed for millions of years, but which by astronomical reckoning is already overdue.

We must take care, however, not to allow the brilliant researches of the molecular biologists during the past decade to dazzle us into thinking that the 'secret of life' has been unveiled by the discovery of the structure of DNA. Much of their work has been based upon investigation of the DNA in bacteria, especially that of *Escherichia coli*, the bacterium that swarms in the human bowel and by contamination on most objects handled by man. The conclusions drawn from the study of bacterial DNA are extrapolated to apply to 'higher' organisms, the eukaryotes which are built up of cells each containing cytoplasm with its organelles and a nucleus enclosed in a membrane in which the DNA is gathered to form the chromosomes.

Research on the DNA of eukaryotes shows that the extrapolation is to some extent justified, but that it must be applied with the greatest caution, as work on the viruses warns. Viruses are minute organisms several orders of magnitude smaller than bacteria; they are scraps of specific DNA or RNA and have been likened to 'footloose genes'. They are able to penetrate into bacteria, or the nuclei of eukaryote cells, and influence their metabolism. The virus particle, or 'virion', has a protein coat containing the 'genome', or nucleic acid complex. When the particle enters a cell it sheds the protein coat, whereupon the genome induces the host DNA to make further complete virions. The result of such penetration, or parasitism, by viruses is often the production of disease in plants and animals, or of destruction in bacteria.

Some viruses contain RNA but no DNA; the Oncornaviruses, which cause cancer, are an important group of RNA viruses. They can make DNA copies of themselves by means of the enzyme 'reverse transcriptase' or 'polymerase'. The DNA thread can interrupt the host DNA thread and insert itself in it; the viral genome then becomes part of the host DNA and cannot be isolated experimentally or demonstrated separately. It may thus persist for years until after some adverse influence or damage to the cell it emerges again as an RNA virus and causes cancerous changes in the cells.

There is, however, an exception: a disease called 'scrapie', which is found mainly in sheep, affects the nervous system with degenerative changes that lead to a lingering death. The external symptoms are dullness and listlessness, and an urge to rub against various objects so that the wool is scraped off, apparently to relieve an itching of the skin. Under the electron microscope the organism of scrapie looks like a virus, but the biochemists have found that it does not contain any DNA or RNA. Here, then, is a minute particulate organism that can replicate itself without the mediation of a nucleo-protein. It has been suggested that a polysaccharide here performs the coding function, for it has long been understood that a self-replicating semi-living particle could be constructed of chemicals other than nucleo-protein. DNA is thus not the only way of writing the genetic code, though in the present state of our knowledge it appears to be the most usual.

It is against this background of a living world constituted as a habitation for DNA that we must examine the interactions between man and wildlife, which are basically inseparable. Man's brain and hands enabled him to draw ahead of the rest of wildlife, and once that had happened he was well justified in thinking that he had been set in dominion over it; it can easily be understood that he thought the world and its wildlife were made specially for his convenience.

Even in palaeolithic times life was not an unending struggle against the environment, for the men of those days had leisure to paint and engrave the walls of their caves, and to whittle bones and stones into representations of animals, in a realistic style that shows startling skill, patience and, presumably, aesthetic appreciation. The urge to find pleasure in making entirely useless things seems to have evolved very early in human prehistory and to have spread throughout the human race. It is strange that an artist is impelled to make a copy – necessarily imperfect – of, say, a landscape when he can see the real thing at any moment if he takes the trouble to look out of the window. Similarly the artists of the old stone age made images of the subjects in wildlife with which they were most familiar, choosing those they liked the best.

Primitive artists probably first executed their works for their own pleasure rather than for exhibition to their companions, though their artifacts were later used, so the archaeologists and anthropologists tell us, in magic rites to promote fertility and success in hunting. There is no reason for thinking that any but a minority of 'cave men' were talented artists, as is shown by the numerous random scribblings and clumsy graffiti that disfigure the walls of their dens – the impulse to such self-expression is with us still and is as strong as it has ever been throughout the intervening millennia.

People with leisure to indulge in the practice of 'art' must have lived in places where wildlife was sufficiently abundant to provide enough food without making their lives one long quest to stay the pangs of hunger. Even the Bushmen of the Kalahari and the aboriginals of Australia, who appear to us to have been living until recently on the lowest level of subsistence, made rock paintings or decorated themselves, their domestic utensils and their weapons with paintings and carvings. Leisure not only gave opportunity for artistic expression and the invention of private property in artifacts, but also found plenty of mischief for idle hands to do. When man's greed led him to covet his neighbour's possessions his aggressiveness, backed by his invention of weapons, started him on the path of war and destruction which he has characteristically followed to the present day.

Man's destructive hand was not turned solely against his human neighbours; it equally affected his environment and its wildlife, for man shares with some of the carnivorous mammals a propensity to kill far more than is enough to satisfy his needs. Beasts of prey do not normally kill more than they can eat – catching their prey is work, and few creatures work more than they must. Sometimes, however, when they have the chance to kill excessively without working hard they indulge in a wild killing spree that seems to be against their own interests, because they are destroying their future supply of food. The depredations of a fox in a hen-house are a well known example, but apart from such artificial situations wild predators sometimes behave in a similar way towards wild prey. Spotted hyaenas in East Africa have been found killing Thompson's gazelles 'wantonly' on a dark misty night when the victims were confused and disorientated so that they did not try to escape – the hyaenas left dead and dying gazelles widely scattered over the plains.

The extinction of certain large mammals, from the mammoth downwards, during the pleistocene epoch is thought by some archaeologists to have been hastened by the overkill inflicted through man destroying more than enough for his immediate needs. Since the beginning of the nineteenth century the fruits of the industrial revolution have given great numbers of people the leisure to indulge their destructive instincts with a huge overkill in the name of sport. In the eighteenth century country landowners like Squire Custance shot a few pheasants, partridges and other game for their own tables and those of their friends, and Parson Woodforde now and then coursed a hare for the same purpose. The parson also enjoyed angling, though always with the object of sending his catch to the cook; but he also supplied himself with fish by netting the local brook – he would no doubt have welcomed a stick of gelignite as a way to bring quick returns, had it then been invented.

The rise of the cult of hunting, shooting and fishing during the nineteenth century as an acceptable way of life for a class of rich people who had nothing to do, drastically affected the wildlife of the countryside. The destruction of vermin in order to make the land carry a population of game far larger than it could naturally support locally exterminated some species and greatly reduced others, with widespread effects on the small mammals and birds. Although the balance was to some extent restored by the reduction of keepering in Great Britain during two world wars the field sports industry is now probably larger than ever, as is shown by the huge rents paid for shoots and fishing beats, and the enormously swollen ranks of sportsmen owing to increased material prosperity and leisure. The vast cost of such amusements in comparison with the value of their product is enshrined in the old pre-inflation jibe at pheasant shooting: 'Up flies a guinea, bang goes tuppence, down comes half-a-crown.'

The industrial revolution attracted labour to the towns from the country after the agricultural slump following the Napoleonic wars. The prospect of comparatively well-paid work and freedom from the tyranny of the squirearchy, under which the rural masses were little better than serfs on the verge of starvation, brought so great an increase in urban populations that the consequent overcrowding and buyer's market in labour pushed the urban lower classes back to conditions of life no better than those they or their fathers had left in the country. A mystique of the simple joys of country life as compared with the horrid conditions of the towns grew up, but it bore little relation to the truth, a false romanticism owing much of its strength to Wordsworth.

'Bright volumes of vapour down Lothbury glide,
And a river flows on through the vale of Cheapside.
Green pastures she views in the midst of the dale,
Down which she so often has tripp'd with her pail;'

But poor Susan's reverie of a 'single small cottage, a nest like a dove's', forgets that it was damp, insanitary and squalid, and that its overcrowded inhabitants lived a life of endless toil in desperate poverty. A cottage that looked pretty enough from the outside was generally no better than a rural slum for those who had to live in it.

In spite of the almost universal illusion about the advantage of living in the country, the flight from the land has gone on with increasing speed right up to the present agricultural revolution in which mechanisation allows one man to do the work once done by twenty – and to earn the wages once earned by twenty. Life in the country is intolerable to most people unless they have the town amenities of public services, domestic gadgetry and transport – and earn their livings in the towns.

The rise in material prosperity which has given nearly every urban family a motor car allows people to escape from overcrowded cities every seventh day to visit the countryside which they have never been taught to appreciate. Their behaviour indulges those destructive instincts that are channeled into sport by others who think themselves members of a superior class; they break down fences, let their dogs chase sheep, root up plants, destroy birds' nests, scatter litter and ordure, get abusive and threaten violence if a farmer remonstrates. By their numbers they trample the land into mire or bare desert, and through ignorance they destroy what they came to enjoy.

Television and the tourist trade are two of the greatest debasers of man's atttitude to wildlife. Not the remotest corner of the world is safe from the degradation brought by those who profit from the craze for tourism, who build concrete matchbox hotels to house their dupes and make charades and burlesques of the lives of the local inhabitants for their amusement. Television gives a superficial view of strange places and their wildlife – it cannot give any depth because the allotted time is too short. Tourism then steps in with the 'package holiday' on which the tourist exerts himself no more than if he were looking at the screen in his own home. He is spared the terrors of making any real contact with the countries he visits, or with the wild or tame life that he sees. All is false and foisted on the trusting 'client' by crafty advertising. You would think from the advertisements that holidays were the sole purpose in life.

The rapid despoiling of the environment by the ease of modern travel and by the population explosion, with the consequent pollution of many kinds, has at last been recognised by a minority of thinking people as a threat to the future both of man and of wildlife. A section of the population in most parts of the world has become anxious to conserve what is left and to prevent further degradation of the earth. Conservation cannot put the clock back, nor can it prevent the increasing occupation and exploitation of the wild places as long as the population explosion goes on; but it can, by modern propaganda methods, urge that we should not repeat past mistakes.

The speed at which the environment is altering is so great that some people think that little can be saved if vigorous measures are not taken at once – in another generation it will be too late. The crux of the matter is that there are too many people; science has found ways of saving the lives of hundreds of thousands of people who a century ago would have died early in life from the effects of numerous diseases that are now regarded as 'conquered'. We have achieved a certain measure of death-control, without a corresponding measure of birth-control; a finite world cannot support a human population that is expanding exponentially and has already 'gone round the bend.'

The subject of the population 'explosion' has attracted much attention during the last decade; one of the latest discussions is the *Blueprint for Survival* (1972) by a team of distinguished ecologists. The writers give an admirable summary of the threat to the environment and its inhabitants – including man himself – from the rapidly multiplying human population, and they make a number of proposals for immediate action to ensure survival. One of the authors points out that on the limited area of the habitable world a population cannot grow indefinitely, and consequently it must stabilise at some point, either of its own volition or be cut down by some disaster such as famine, epidemic disease, or war. He then goes on, 'Since no sane society would choose the latter course, it must choose to stabilise', and pursues his argument about what should be done on the assumption that societies are sane. If one accepts this assumption one cannot but agree that nearly everything suggested is sensible and necessary. The assumption of the sanity of societies is, however, the fatal flaw in the argument, the great fallacy that brings all to nought.

Look at the scenes of hooliganism at football matches in Britain, and of the violence and bullying that too often go with 'industrial action'; think of the wanton destruction wrought by vandalism. On a larger scale consider what has happened in Ireland during the last five years, and in Viet Nam during the last ten; think of all the atomic missiles lined up on each side of the Atlantic aimed at all the great cities of two continents. Think of the economic system that demands expansion in order to survive, turning out millions of pounds worth of consumer durables and consumables that are not necessary, and spending nearly their value in advertising to persuade people to want them. Think of the property developers and of what they have done to the shores of the Mediterranean, the Caribbean Islands, and many other once pleasant spots. Societies are not sane in the sense meant by the writers of Blueprint for Survival. The great majority of people does not want to know about the population explosion and all that it implies, and could not form a sane judgment if it did.

Nevertheless the minority who know the dangers ahead bravely soldier on, preaching their gospel of population control and conservation of the environment in the face of indifference and of obstruction by vested interests. In spite of difficulties they are making some progress – their most sympathetic hearers are the women of all nations to whom they bring the prospect of release from the recurring burden of unwanted pregnancy. Time is against them, yet the work of such organisations as the International Planned Parenthood Association is making itself felt all over the world. But until governments take courage to face the problem and impose laws that would have been intolerable a generation ago and would be unpopular now, the efforts of voluntary bodies cannot bear fruit before it is too late, though they are laying a foundation for official action. If that action does not come soon the alternatives listed by the writers of *Blueprint for Survival* will solve the problem in the most dreadful ways.

In the meantime, assuming that a substantial reduction in the rate of world population growth is attained, what can be done to prevent further deterioration of the environment? In the first place we shall have to learn to refrain from polluting our environment with poisonous or otherwise undesirable substances that cannot be quickly made harmless by natural processes. As an example, consider the chlorinated hydrocarbons which have brought immense benefit to millions and added to the population problem by controlling the insect vectors of disease and the insects destructive to food production, but which have had unforeseen harmful side-effects on the environment. The herbicides, too, have had a similar double action. Technology is constantly searching for new chemicals which are hoped to have the desired control without unwanted side-effects. Until they are found we must face the fact that large amounts of deleterious substances will continue to be released into the environment because their designed effects are so valuable. The best that can be hoped is that education will lead to their more careful and discriminate use. On the other hand pollution of the seas with oil, and of rivers with sewage and industrial wastes, could be enormously reduced – we have the knowledge of how to do it but lack the desire to spend the large sums of money needed. The problem has crept up as gradually as the population has grown – the environment can purify a certain amount of pollution, but is overwhelmed when more is dumped into it than it can deal with.

Pollution with discarded objects made of plastics is in the main a different matter, it disfigures the environment rather than poisons it, though small amounts of the plasticisers used in their manufacture are thought to be possible dangerous contaminants of some seas. The fault of plastics is their lightness, so that the winds blow them about the landscape, and the waters cast them ashore where they persist as unsightly garbage because they are not 'biodegradable'. Research is now striving to find plastics that are destroyed by micro-organisms or by the ultra-violet rays of sunlight – an apparently crazy quest, because the whole point of plastics, and their usefulness, is due to their inert resistance to destruction and their lightness. It is peculiar that although we have been contaminating the environment with discarded containers made of glass for several hundred years no public anxiety has been expressed; but glass, being much heavier, soon gets buried in the earth or sinks to the bottom of fresh and salt waters – the bottom of the Antlantic beneath the shipping lanes must be nearly paved with empty bottles embedded in the ooze.

Polluting the environment by dumping garbage, old bicycles and bedsteads in ponds and rivers, where they are more or less out of sight if the waters completely cover them, is nothing new. 'And the priests went into the inner part of the hosue of the Lord, to cleanse it, and brought out all the uncleanness that they found in the temple of the Lord into the court of the house of the Lord. And the Levites took it, to carry it out abroad into the brook Kidron.' II Chronicles, 29, 16.

Conservation is not preservation, nothing is static or ever has been, and we must accept change whether we like it or not. Conservation consists of clearing up our mess as we go along, and not dumping it out of sight. It also means refraining from destructive exploitation of replaceable natural resources, and in salvaging and recycling those that are not; though the exhaustion of fossil fuels will present a problem that will have to be solved by other means. If we are to avoid destroying natural resources we shall have to manage them, just as agriculture manages most of western Europe outside the towns – an enterprise that will presumably apply to the whole surface of the world in the long run. Many sorts of wildlife will no doubt be able to live in peaceful co-existence with man – even in Great Britain we have a huge flora and fauna in spite of the centuries-long settlement of the country. On the other hand those species that cannot adapt, if they are not exterminated, will be restricted to parks and reserves where the environment cannot be wholly natural if only because it has man-made boundaries.

This state of affairs will be accepted as right and natural by future generations, just as we accept the artificial environment of civilised Europe. Success in keeping the world a pleasant place to live in depends primarily upon educating the population to limit its numbers and secondarily upon teaching it to understand and respect the fragility of the environment so as to avoid destroying it.

In the chapters that follow we shall look in more detail at man's relation ship to different forms of wildlife, consider how he has treated them in the past, and how he may perhaps alter his attitude towards them in the future.

2 MAN, FIRE, PLANTS AND TOPOGRAPHY

It is certain that for the greater part of his existence on earth man was a hunter of animals and a gatherer of plant products, for it was not until some ten thousand years ago that he invented agriculture and domesticated a few kinds of animals. During the tens of thousands of years in the hunter-gatherer stage of his evolution man was still part of wildlife – just another species living in small groups based upon the family; each occupying a limited territory. The effect of his activities upon the rest of wildlife was probably negligible, for he could not have become a dominant species before families united into clans and tribes in which the cooperation of individuals, in a species physically weak compared with other animals of smilar size, brought the large animals within reach as regular prey. The coherence of the family was probably in large part due to the long period of childhood with a consequent necessity for prolonged parental care, which strengthened what ethologists call the 'pair bond'. The evolution of language also strengthened this bond, and made possible the cooperation of family groups for common interests.

Man early learnt the use of fire, for wildfire is common enough in the tropics, and has been so for millions of years, as is shown by the microscopic particles of carbon still retaining the characteristic form of particles derived from forest fires, found in the bottom deposits of the oceans. A natural bush- or savannah-fire concentrates the fleeing wild animals in its path so that they are highly vulnerable to primitive hunters; once man had captured fire he was able to use it for driving his prey in the absence of wildfire. After the invention of agriculture the use of fire for clearing forest land of trees and bush became a widespread way by which man brought about changes in wildlife, as discussed below. It is doubtful whether man used fire for cooking in his early hunter-gatherer stage, for raw meat is both palatable and more easily digested than cooked – do not many people even now prefer their steak 'rare', or 'saignant' as the realistic French express it? It was not until foods containing starch became common that cookery was needed to burst the coats of the starch-grains and make the contents available to the digestive enzymes.

Cooking meat is a luxury rather than a necessity; the gastronomic value of cooked meat was no doubt discovered accidentally much as Charles Lamb's Chinese boy Bo-bo discovered the virtue of roast pork, but why we enjoy the savour of so unnatural a food as fire-scorched meat remains an unanswered question – is the taste innate or acquired? Does the fact that the petrels, the oceanic sea-birds which almost alone among birds have a well-developed sense of smell, can be lured from great distances by the odour of frying pork or blubber, a delicacy entirely unnatural to them, suggest that it may be innate? Preparing food by boiling, on the other hand, had to wait the invention of vessels either of pottery or fine basket work, or the lined cooking hole in the ground heated with pot boilers – stones heated in the fire and plunged into the water – a method also used with soft fired pots. Boiling some kinds of flesh certainly makes it more edible, for it softens hard or keratinous materials such as tendons and skin, and tough muscles so that they can be chewed and digested. Culinary art must have been born when herbs and other vegetable matter were added to the pot to make a stew.

Before the invention of cooking the institution of the camp-fire as a source of warmth and protection must have been one of the earliest uses of fire by primitive man. Without it, life outside the tropics is at best uncomfortable, and after sunset the flames lighten the darkness to help ward off prowling predators and the imagined terrors of the night. The Romans called the hearth *focus*, a word we have adopted to mean the centre of attraction or point of concentration, and held the hearth equal in value to the altar in their moral slogans, as in Cicero's 'pro aris et focis'. The place of the hearth, the old camp-fire, has held its importance as the centre of family life to this day; when real fires are replaced by electric heaters we make fake logs and coals to glow in imitation of the real thing as the focus of interest, though with the spreading use of domestic central heating the focus is moving from the fireside to the television box, almost the sole means by which the urbanised masses have a faint vicarious contact with wildlife. Our respect for the camp-fire is preserved in popular saying and song as in 'Keep the home fires burning' and the symbolism of the Olympic flame – and what about those gas jets burning under the Arc de Triomphe in Paris, and in Arlington Cemetery, Washington, D.C.?

Until man found how to make fire when he wanted it, keeping the home fire burning must have been an important duty. We think of fire-making with bow-drill or flint and iron-stone as laboriously primitive, but until about 150 years ago our only way of making fire was with the tinderbox, to catch a spark struck by flint from steel on to charred rags and rotten touchwood or amadou, the dried flesh of a polyporus fungus. Only fifty years ago the 'strike-a-light' flint and steel were popular toys used by boys on dark evenings.

Fire, which has been one of the central points of human civilisation, though we now take it so much for granted that we have forgotten its importance until there is a strike by miners or electricians, was the first means by which early man had any large effect on wildlife. With a few sparks he could start a forest or grass fire that roared away out of control and changed the environment in a way that years of labour would be unable to effect; it gives power far greater than the strength of individuals. In his hunter-gatherer stage man probaly found that fire could be used for more than concentrating his prey by destructive burning; he realised it could also be used constructively when he noticed that a grass fire was soon followed by a flush of new growth that attracted grazing animals to recently burnt areas. By thus altering the environment he was able the more easily to exploit wildlife for his own ends. We do not know when man's activities began thus to affect wildlife; it may well have been before the appearance of *Homo sapiens*, for earlier species of man living in Asia were using fire some three hundred thousand years ago, and something less than a quarter of a million years later *H. habilis* was using fire in Africa.

If the early impact of man on his environment was thus directed to the procurement of animals for food, the main visible effect of it was necessarily upon plant ecology. A patch of forest destroyed by fire regenerates as secondary forest with a different species-composition from that of the burnt primaeval forest, and it may take a century or more for the succession of species to attain to a climax forest indistinguishable from the original. Apart from burning, however, it was not until the beginning of agriculture some eight or nine thousand years ago in western Asia that human activities started affecting wildlife seriously.

The advent of agriculture forced man to live in settlements which were either permanent or, if seasonal, lasted long enough for him to sow and harvest a crop, or several successive ones. He no longer wandered through his territory as a nomad leaving scars on the environment which for the most part quickly disappeared from it. Settlements for agriculture needed clearings for fields, and dwellings for people, the first step towards civilisation and the urbanisation of today. The comparative security of settlements, and the enhanced assurance of enough food given by agriculture, allowed an increase in the size of the population, which grew with positive feed-back as new skills were learnt and the division of labour was gradually adopted.

Agriculture, the cultivation of cereal crops, must have come almost imperceptibly out of the hunter-gatherer way of life among people who had long been gathering the seeds of wild grasses for food. Some cultivated cereals, such as wheat and barley, are polyploids – the number of chromosomes in their cells is a multiple of the number in wild species; modern varieties are produced by scientific plant-breeders using extremely complex techniques of crossing and selection. The first cultivated cereals must have been selected and segregated from natural hybrids of wild species that gave larger and more abundant grains. The use of such improved hybrids made agriculture necessary because they cannot survive in the wild; although cereal crops cover many millions of acres of the earth's surface they do not escape to grow in the wild. If they were able to survive without man's help they would dominate the flora of many countries, but when superceded varieties are allowed to go out of cultivation they become extinct. The successes of the plant breeders are astonishing; the advent of the barley variety named Proctor a few years ago changed Great Britain from a barley-importing country to an exporter of her surplus. The inventor was rewarded with a prize of £1,000, equivalent to three months' salary of a medium-grade business executive.

The cultivation of cereals may not have been the beginning of agriculture, for it is possible that other plants such as pulses were taken into cultivation first; it has also been suggested that the cultivation of edible roots in tropical countries may have preceded the use of cereals. Be that as it may, it is certain that the first culture of cereal crops in western Asia some nine or ten thousand years ago was a landmark in the evolution of western civilisation, so that large city communities, with all the implications for the development of the useful arts and sciences, could arise, dependent upon the surrounding agricultural lands for their food.

Under the climactic conditions of western Asia, lands cleared of their natural cover for the purposes of agriculture lose water and start becoming arid. It is something of a paradox that plant cover conserves soil moisture because the amount of water evaporated into the air by transpiration from the leaves is immense – a plant acts almost as a wick. On the other hand a covering of vegetation protects the soil from direct loss of water through evaporation under solar radiation, and moreover foliage, particularly of herbage, facilitates the deposition of dew when the temperature falls at night. Soil without cover quickly loses water under the influence of sun and wind, the moisture beneath being drawn to the surface by capillary action, to break up which every gardener knows that he must keep the surface well stirred with the hoe in dry weather.

When land gets dry water can be brought to it by irrigation, an art possibly learned in Egypt where the annual inundation by the rising Nile provides both natural irrigation and a top-dressing of fertilizing silt. Irrigation channels dug to bring water from streams and rivers to fields of crops leave long-lasting scars on the topography, and provide new local environments for wildlife. Indeed, the permanence of even minor man-made earth disturbances is surprising: ditches, trackways, field boundaries and the remains of settlements can be traced from the air thousands of years after they have been obliterated from surface view – ground once disturbed generally carries plant communities differing in some degree from those of adjacent areas. The difference, though small, has a far-reaching effect on the complex web of wildlife, from soil bacteria to higher plants, and from small invertebrates to birds and mammals. Larger works such as terracing, which has been practised from early times, and the construction of rail- and motorways and dams in modern times, have correspondingly greater effects on wildlife, and will probably be visible for thousands of years until they are destroyed by another glaciation.

In regions where rainfall is sparse or markedly seasonal the removal of natural herbage to make way for cereal crops can lead to disaster if irrigation is neglected. The soil particles, no longer held by the roots of herbage, are blown away by the wind, leaving 'dust-bowls', or are washed away by heavy rains which run off the surface instead of being held by the herbage so that the water can soak into the soil. In ancient times much of the north African desert was man-made in this way – the lands bordering the southern shores of the Mediterranean sea were once the granary of the Roman empire. Some of the deserts of central Asia were similarly once fertile, and were inhabited by large populations whose ruined cities now lie deep under arid sands. In modern times the dust-bowls of parts of the United States were made by ploughing for the cultivation of cereals, which exhausted the soil's fertility. Even in England the rich black powdery soil of the Fens is blown away in unfavourable seasons if dry windy weather comes before growing crops are established, the young plants as well as the surface soil being carried off by the gales. In addition to the plough man's domestic animals are often powerful influences in making deserts; goats are particularly destructive for, being browsers, they prevent the regeneration of tree cover and reduce once fertile regions to arid maquis.

In tropical countries 'slash and burn' agriculture is practised by semi-nomadic people, who cultivate small clearings for a few years until the soil fertility is exhausted. They then move on to make a new clearing, and the forest ecology is seriously altered if the population increases above a low level. Abandoned clearings regenerate if given enough time to lie fallow, but if they are reoccupied too soon the thin layer of top soil is lost, and the action of the sun on the iron and manganese in the underlying soil causes laterisation with the formation of hard-pan, which is impervious to water so that excessive run-off occurs, vegetation cannot find a root-hold, and semi-desert or useless scrublands take the place of the forest.

The planting of forests has a drastic effect upon ecology and wildlife as great as the destruction of natural ones, especially when the planted species differ from those of the native flora, as in replacing deciduous woodland with evergreen conifers. In Great Britain many people are prejudiced against afforestation with conifers in order to obtain mature timber as quickly as possible, but conifer forests, though widely different in character from deciduous ones, are equally pleasing aesthetically, and as interesting ecologically.

Although until recently it has been impossible to reclaim a man-made desert, which once made seems permanent, land can be won by draining marshes, but it must be constantly guarded – if vigilance is relaxed it will soon be lost to the waters. If embankments and drainage channels, sluices, and pumps are damaged the labour of years may be lost in a few hours. It is surprising that the winning of land by the drainage of marshes has been practised for so many centuries, because when it started the human populations were not so large that they were driven to it by land-hunger. It is probable that the early attempts at drainage, as for example the beginning of the draining of the fens in East Anglia by the Romans, were made for political reasons to destroy the inaccessible habitats of recalcitrant subject peoples. Whatever the reason, no more striking change in ecology can be seen than the immense fields of crops flourishing where great meres, peat mosses, and reed beds once covered the countryside, and supported a fauna and flora totally different from those now found there.

The amazing topographical engineering of the Dutch in adding hundreds of square miles to their native land is one of the wonders of the world, though much of the work wins back once dry land that was lost by inundation from the sea in mediaeval times. Yet even with the aid of the most modern machinery the Dutch rely on biological means for part of their reclamation, and bring wildlife to their aid. In the Ijssel Meer, formerly the Zuider Zee, they build an embankment round each new polder, and then pump out the water leaving a vast extent of oozy mudflat. In order to dry out the new land they plant huge beds of the common reed, *Phragmites communis*, a plant that can grow in salty soil. The transpiration of the reeds evaporates great volumes of water so that in a few years the ooze is consolidated, and fertilised by the decaying roots and plant remains. As the land dries and comes into agricultural use many species of plants appear in addition to those cultivated and to the trees planted for shelter belts. After the plants come the animals, from invertebrates to small mammals and birds, including many kestrels which nest in the boxes on poles that have been erected for their use in order to attract them to the polders, where they prey upon the mice and voles.

The opposite process to making a new dry-land habitat is seen in East Anglia where the broads artificially make a new water habitat. The broads were long thought to have been formed by natural means in the courses of the meandering rivers of the low-lying lands of Norfolk and Suffolk, but modern study has shown that they are the water-filled holes left by mediaeval and earlier peat diggers – the remains of ancient turbaries. The whole district has been given a man-made ecology as striking as that produced by draining the fens or destroying the Scotch forests, so that it now has a characteristic flora and fauna.

Enlisting the help of plants to alter the topography of the land has been carried on for centuries. The planting of marram grass, which propagates by sending out long creeping underground stems, for consolidating loose sand-dunes has long been practised with success. The spiky leaves, rolled into cylinders, protect the stomata and reduce loss of water by transpiration, and the plants are tolerant of salt in the sand. The roots hold the sand grains, and the tufts of grass hold debris, so that humus begins to form and other plants can gain a footing, with the result that former wind-blown dunes are covered with turf and bushes. Marram has thus been a great help in strengthening coastal defences to prevent the sea flooding low-lying coastal lands, and has laid the foundation for new habitats for wildlife.

Such use of suitable plants, however, sometimes has unexpected and unwanted effects. Cord grasses of the genus *Spartina* grow on the tidal mud flats of the coast and estuaries on both sides of the Atlantic. About seventy years ago the American species *S. alterniflora* was accidentally introduced onto the mudflats of Southampton Water, where it became established and where it crossed with the English species *S. maritima.* The hybrids were not only stout, strong-growing plants showing 'hybrid vigour', but bred true, unlike most F_1 hybrids. At fertilisation they had become polyploids with the nucleus of each cell containing the full number of chromosomes of each parent instead of the normal half. Indeed they could not have hybridised otherwise, because *S. alterniflora* has 70 chromosomes whereas *S. maritima* has 56, so that the pairing of chromosomes from each parent, normal in fertilisation, could not take place. Instead of the cross being sterile through the incompatible number of chromosomes, by some odd chance the chromosomes from each parent divided and paired in a single cell, with the result that *S. townsendi* has 126 chromosomes, and is a new species differing from both the parent species. It soon started colonising the mud flats of Southampton Water, ousting the native species and forming such dense meadows that they trapped the water-borne silt brought with every high tide, and held it so that the level of the mud flats rose. The consolidation was so rapid that oozy mud-flats were soon converted into dry land. When the land-reclaiming ability of *S. townsendi* was appreciated the new species was widely used by those concerned with coastal defences in many parts of the world. In some places the cord grass has got out of hand and has encroached on harbours that are wanted as open water and not dry land; those who introduced it find that it is much more difficult to eradicate than to plant it.

The extent of man's dependence upon grasses is surprising – reeds, marram and cord grass are no less grasses than the cereals of western civilisation. Grasses, however, have been cultivated for human needs from prehistoric times in other parts of the world; rice, a water-grass native to eastern Asia, is cultivated throughout the damper parts of the tropics and warm temperate zones to form the staple diet of half the world's population. Other wild swamp rices occur in Africa and America. Sugar cane, another tropical Asian grass, is universal in the tropics, as are bamboos, the shoots of which are used for food and the stems for many constructional purposes. Millet, sorghum and other grasses of the warmer parts of the Old World also provide edible grain. In the New World man found and domesticated another grass, maize or Indian corn, the only cereal native to the Americas.

Agriculture must therefore have been invented many times by people in various parts of the world, not only for growing cereals but for the cultivation of roots such as yams among the peoples of the Pacific islands who had no cereals. Whether the cultivation of roots such as mandioca and potatoes in America came before or after the cultivation of maize has not been determined beyond doubt. Though cereals form the staff of life for much of humanity, an immense number of herbs and trees yielding edible fruits, roots, foliage or nuts have been brought into cultivation to add variety or luxury to the diet.

The plants producing fibres are second in importance only to those producing food – flax, hemp, cotton, sisal and others, of which flax has probably been cultivated the longest. The cultivation of fibres for anything but the simplest fastenings implies that the users must have reached a high degree of technical civilisation and have invented instruments for preparing the fibre and for spinning cordage, or threads for weaving into textiles.

The use of india-rubber provides an example in modern times of the transition from gathering wild products to cultivation. Rubber is produced from the latex of many species of trees and lianas of the tropical forest, and until less than a hundred and fifty years ago it was obtained entirely from wild plants. The true rubber tree, which produces Pará rubber, was taken into cultivation from Brazil, its native land, to Malaya and the East Indies, where the plantations produce some 90 per cent of the natural rubber crop of the world.

The great variety of vegetable drugs formerly in use were originally gathered from the wild, though many were later brought into cultivation in herb gardens or larger plantations. The majority are now rejected by medical practice, though some – or their synthetic equivalents – are still in use. The narcotics such as tobacco, hemp and the poppy are grown in large crops for man's solace and enjoyment apart from their use in medicine, though narcotic containing fungi such as the fly agaric, which is used to make an intoxicating drink in parts of Asia, are only found wild. The use of the yeast fungus for producing carbon-dioxide gas to leaven bread, and for fermenting sugar to make alcoholic drinks, began in the early stages of civilisation – yeast is probably the most universally-prized species in the whole vegetable kingdom.

Seaweed is a crop that is still gathered wild in Western Europe, from a plant that has never been brought into cultivation. For hundreds of years the sea wrack, consisting of various seaweeds cast ashore often in great quantities by the waves, has been gathered for several purposes. The simplest use is as an agricultural mulch for top-dressing the land; in the west of Ireland for example, where it is strewn over the fields, the dark green of the grass growing where bits have fallen and decayed contrasts strongly with the paler grass in the spaces between. The most important use, however, was in kelp-burning to produce a calcined ash rich in potash that was used in the manufacture of soap and glass. It was later valued as a source of iodine, but these industrial uses are now obsolete.

'Kelp' was the name applied both to the ash and to the large brown laminarian seaweeds with leathery fronds that grow in the zone below the level of low tides and, when detached by storms, often form the bulk of the wrack washed ashore. In recent years seaweeds have again been brought into use for making horticultural fertiliser, and on a large scale for the extraction of alginate, a substance widely used in the food-processing industry; and to a lesser extent for other purposes. Special gear has been invented for cutting and gathering kelp from the dense beds it forms on the sea bottom in coastal waters. The beds regenerate two or three years after being cut, so it would perhaps be a waste of effort to try to cultivate kelp artificially.

Wild seaweeds of other kinds are also gathered for various uses, such as the edible dulse of Scotland and the laver which is eaten in South Wales as 'laver bread', or the agar-agar prepared from a red seaweed and used universally by microbiologists as a culture medium for growing bacteria. The gathering of these wild crops, especially the mowing of kelp beds, must have a considerable effect on the local ecology even though it may be only temporary. The gathering of sea-wrack for kelp-burning removed an entire micro-environment, for the decaying weed is the habitat of great numbers of invertebrates which, if undisturbed, are the mediators of the cycle by which the tissues of the weeds are broken down and returned to the sea.

On a global scale timber from forest trees is still largely a wild crop, though the deliberate replanting of clear-felled lands, and the afforestation of degraded soils, is increasing in importance – and is no more than common sense. The destruction of the forest has been going on for thousands of years, not only in order to get timber but in the course of settlement and husbandry. It has recently been pointed out that the blanket-bogs of Wales were produced as far back as neolithic times by man and his domestic animals interfering with upland woods. The opening up of the forest canopy and eutrophication, together with a deteriorating climate led to soil leaching and the formation of bog. Neolithic man could not understand the damage he was doing to his environment, but the modern devastation of forest by the wood-pulp industry cannot be justified unless forests are replanted and the pulp used economically. During the Second World War everything of importance could be printed on single folded-sheet newspapers, so are we really justified in squandering vast areas of forest in order to disseminate ephemeral rubbish to titillate the baser emotions?

The cultivation of timber crops, like that of all others, is a fight with wildlife. Every plant provides food or lodging, or both, for a vast range of vegetable and animal parasites and predators, and has to compete with rivals for nourishment and living room. In the wild the predators and parasites have to work for their livings, seeking out or drifting in air currents upon their victims, sometimes successfully but often in vain. In the wild the victims and the aggressors generally reach a balance, so that neither group is totally destroyed. But when man interferes with the relationship by growing crops of various plants in pure stands, and eliminating the competition of weeds, he provides a feast on to which all the beasts of the field and the fowls of the air descend with joy, and on which the insidious bacteria and viruses batten out of sight, together with the moulds and fungi – the 'blasting or mildew, locusts or caterpillars' named by Solomon when he prayed as he knelt on his golden scaffold (II Chron. 6, 28).

The endless battle with animal pests has been waged by man with everything from traps, scarecrows, magic and watch-boys with rattles, to carbide bangers, modern organo-chlorine chemicals and nerve gases. The battle has never resulted in anything better than a draw until science came to the help of empiricism in the second half of the nineteenth century; but after some notable triumphs science has now become almost too successful. Apart from the application of chemicals to kill pests, science had some early successes with biological control. One of the best known examples is the introduction of the *Cactoblastis* moth into Australia, where the introduced prickly pear *Opuntia* was running riot and occupying vast areas wanted for agriculture. The insect – brought, like the plant, from America – was released on the cacti, destroying the pest species wholesale and reducing the nuisance to negligible size.

A less successful attempt at biological control was the introduction of mongooses from Asia into some of the West Indian islands to destroy the rats that were devastating the fields of sugar-cane. The mongooses not only fed on the rats but also attacked the native mammals and birds, bringing some species to the verge of extinction. The organo-chlorine pesticides, which seemed to be the perfect answer to the problem of pest control when they were first introduced, have produced serious unwanted side-effects in killing useful species and polluting the environment in general. Science cannot be blamed for this; it has been brought about by the often excessive use of pesticides by people ignorant of their side-effects, and encouraged by chemical manufacturers interested only in boosting profits and the tonnage of their annual production.

Now that the scale of pollution produced by pesticides is known there is growing support for those who advocate moderation in their use – some compounds are prohibited in certain countries. This has refocused attention on methods of biological control which may be hoped to be less harmful to the environment. The discovery of the pheromones, the chemical substances that attract the opposite sexes of insects to each other when released into the air in minute quantities, may be adapted to lure insect pests to their death. The pheromones are distinct for each species and thus may be expected to have no unwanted side-effects. Another promising attack on pests now in its experimental stages is to rear large numbers of males of the insect pests and to release them after they have been sterilised by x-radiation. The males seek out the females which stop releasing their pheromone after copulation, even with a sterile male, and refuse further mates, so that they lay infertile eggs.

The experience of unintended pollution with pesticides shows that the most important property to be sought in new compounds is that they should have specific action on a particular pest, but be completely harmless to all else. Biological control, if it can be devised, may well be able to meet this requirement, but considering the lesson of the mongoose in the West Indies and many similar experiences elsewhere, the need for careful study of possible unwanted consequences is evident. On the other hand some chemical pesticides have been unqualified successes – the use of the anti-coagulant coumarin compund known as 'warfarin' for the control of rats is now well known. Warfarin is far less toxic to other rodents such as mice, so that it is almost specific to rats, and has reduced the once serious problem of rat infestation to a minor matter. Sooner or later, however, a genetic combination giving immunity to warfarin was bound to arise in the rat population and, once it appeared, to spread rapidly because the use of warfarin in effect selected the immune rats for survival. Several such strains have arisen, and it has been calculated that the immune strains will have replaced the susceptible ones throughout Great Britain by the 1980s. It has even been suggested that some immune strains are now drug-dependent upon warfarin, so that the way to kill them is to discontinue the use of poison! Science has been searching for alternatives to warfarin since the immunity first appeared in limited areas some years ago, and in February 1973 the discovery of a new compound at present known as UK 786 was announced as a possible species-specific poison for rats which is comparatively harmless to other animals.

The plant pesticides, the selective weedkillers, based upon excessive doses of plant-growth hormones which cause a burst of proliferation followed by death, are not polluters of the environment, though their effects on the composition of the flora are great. They destroy some species of broad-leaved weeds but have no effect on the narrow-leaved cereals and grasses to which they are applied. Fields of cereals are now far cleaner than farmers of twenty-six years ago could have imagined – and no longer do we see them aflame with scarlet poppies when the ears are ripening; it is perhaps a pity that the unwanted weeds of cultivation are so often aesthetically pleasing. Chemical fungicides, too, have been widely and successfully used in combating the many moulds and other microscopic fungi that spoil all kinds of crops. There has been no outcry against any pollution of the environment or unwanted side-effects that they may cause, perhaps because they bring none, or perhaps because fungi are little esteemed and are looked upon as poisonous or not nice by most people. On the other hand, one may wonder whether they are implicated in the almost universal disappearance of the field mushroom, once so abundant in our autumn meadows, and so superior in taste to the cultivated strain – or is the extinction due to the lack of horse manure on the pastures?

The plant-breeders and geneticists have made great contributions to the control of plant diseases caused by fungi, bacteria and, above all, viruses, in breeding strains immune to specific infections. Their work, however, is without end, for new strains of infection are continually appearing to challenge their efforts. Breeding immune strains of plants is naturally confined to cultivated crops, and appears to have had little effect upon wildlife.

Man's husbandry, as well as his agriculture, has equally profound effects in modifying the environment and its wild flora and fauna. The introduction of domestic animals into islands has often had disastrous effects, although done from the best of motives to provide food for castaways. Goats and pigs introduced deliberately, as well as cats and rats introduced accidentally, have destroyed native vegetation and brought about the extermination of many plants and animals. On a larger scale the introduction of rabbits into Australia, followed by unsuccessful attempts at biological control by introducing foxes, is the classic warning against irresponsible interference with the environment. The use of myxomatosis, a fatal pox disease specific to rabbits, for killing millions of the pests was a triumph of biological control, but, as with rats and warfarin, strains immune or resistant to the disease soon appeared. Those concerned with food production have, however, to contend not only with wild pests but with a wilder human illogicality – when myxomatosis appeared in England in 1953 the Government, which had in the past spent huge sums of money on fruitless attempts to control the rabbit population, actually tried to contain the epidemic and stamp out the disease. Myxomatosis, and delighted farmers, fortunately defeated such ridiculous antics. Had myxomatosis been a disease of rats no-one would have objected to it.

The influence of domestic animals on the environment is greater than is generally realised. Much of the maquis country of the Mediterranean region owes its arid character to the destruction of all but the toughest vegetation by goats; equally striking but less known is the way in which the character of the whole countryside of Great Britain has been moulded by sheep. Apart from the growing of cereals, which was for long protected by the Corn Laws, the raising of sheep was by far the most important part of British agriculture from mediaeval times until well into the nineteenth century. The magnificent 'wool churches' of East Anglia, the Cotswolds and many other parts of the country, built by prosperous wool merchants and cloth weavers, are witness to the importance of sheep in the economy, as were the drastic laws forbidding the export of sheep and unmanufactured wool. The vegetation of the whole country is sheep-influenced, from the mountain tops to the prés-salés of the lowland shores. The forests of the hills were destroyed by sheep, which nibble seedlings so that natural regeneration cannot occur, producing peat bogs on the mountains and thin shallow turf on the downs and wolds. Where sheep – and rabbits – are denied access, seedling trees soon appear and establish woodlands within twenty years. The entire British landscape is thus artificial and made by man through the mediation of his sheep, which by their modification of the vegetation have, until the industrial revolution, determined the nature and distribution of its wildlife.

3 EXPLOITING WILDLIFE IN THE SEA

The sea has ever provided a source of food for man. From mesolithic days, and doubtless for thousands of years before, shellfish were a staff of life to coast-dwellers before the invention of agriculture, to the hunters and gatherers who lived upon what they could find. The shell-heaps or kitchen middens on the coast of Denmark, some nearly a quarter of a mile long, fifty feet wide, and ten to twenty feet high, dating from the early Neolithic, are probably the accumulation from centuries of shell-fish gathering. Similar mounds are found on the coasts of North and South America and of east Asia – indeed, the natives of Tierra del Fuego, whose culture was still that of the stone age until the arrival of Europeans, were adding to their shell middens almost into the present century. These Indians, protected from a harsh climate only by cloaks of animal skins and flimsy sleeping shelters at night, were hunter-gatherers in whose diet raw shellfish gathered from the beach were an important part.

Skara Brae, a pictish village in Orkney, was buried to the tops of the hut walls in its midden, so that it was almost underground. Although its inhabitants were living in the bronze age they knew nothing of metals, for their culture was a relic of the stone age, cut off from the rest of the world on a remote island. They had no cereals, textiles or wood, but they did have domestic animals; so their diet consisted of mutton and beef supplemented with great quantities of shellfish, mainly limpets. They probably lived in more comfort than the Tierra del Fuegian Indians for they had peat fires in their stone huts beneath the midden, but the stinking squalor in the houses, as shown by the rubbish they had not thrown out, must have been appalling.

In our own times even civilised man has produced waste-heaps of shell mounds when circumstances have forced him to rely on shellfish for the main part of his diet. In 1916 ninety-eight British sailors from a torpedoed ship, held prisoners by the Arabs in Libya for five months, were given such meagre rations that they kept alive by eating enormous quantities of land snails. The empty shells, bleached in the sun, made huge midden-mounds. In civilised countries, however, shellfish form a supplement, often a luxury, to the diet, rather than a staple component of it. Nevertheless, apart from the oyster – so highly priced as to be the prerogative of the rich – great quantities of cockles, mussels, clams, whelks, periwinkles and other molluscs are eaten by the masses as a treat rather than as serious meals.

Cockles are generally cooked and shaken out of their shells before being sent to market, so that the fishermen accumulate large mounds of shells which were formerly burnt for lime. Cockles are gathered by raking them out of the sand in which they lie buried at low tide; the once extensive cockle fisheries of the sandy shores of South Wales have been severely damaged by the depredations of huge flocks of the oyster-catcher – a bird protected by legislation enacted by people who probably look upon cockles as a repulsive indulgence of the lower classes, which they themselves have never deigned to taste. Here wildlife valued by one part of the people is threatened by wildlife favoured by another, which for sentimental reasons can be complacent about the damage to the livelihood of their neighbours. In the hard times after the First World War poor people in the south-west of England earned a few pence by gathering limpets for food from the rocks at low tide, though limpets, which they called 'rock-beef', do not make an attractive meal. The muscular foot of the animal is extremely tough, the black visceral mass is not beautiful, and the long radula, the rasping tongue set with numerous minute prickly teeth, is a menace to the tonsils.

An obscure backwater of mollusc-eating flowed on for several centuries among the glass-blowers of Bristol, who ate common garden snails because they believed they strengthened the lungs. Before 1914 dilapidated old men were to be seen shuffling round the suburbs of Bristol carrying small, dripping-wet sacks full of what they called wall-fish, the garden snail *Helix aspersa*, which they gathered from the gardens of unoccupied houses. A medicine, too, was made from snails as a cure for consumption: a couple of dozen or so threaded alive on a string were hung above a plate sprinkled with sugar on to which they dripped their slime. The sugar-slime mixture was swallowed to fortify the lungs and drive out the tuberculosis, perhaps because of the resemblance of snail mucus to phthisic phlegm, invoked the "doctrine of signatures".

Kitchen middens are important to the archaeologist because they contain many things besides shells and bones that show how the mound-makers lived. They interest the biologist not only because they sometimes show how the fauna has changed since they were made, but particularly because they show that in spite of the huge harvests gathered from the sea, neolithic man was not seriously affecting the environment adversely – he was not cropping the stocks beyond their sustainable yield. The human population was not large enough to deplete the supply during several centuries, and a balance was held between prey and predator.

People living at the waterside, whether of sea, lake or river, are bound to be familiar with fish – even the most primitive gatherer of shellfish can often grab a small fish from a rock pool, and would surely want the big one that got away. It is impossible to know which of the many devices for capturing fish was first invented – methods differ in different parts of the world, so that primitive inventions in one place may have been later inventions in others. Palaeolithic man has left us beautiful engraved pictures of salmon, but not of the method by which he caught them. In the days when human populations were small, salmon populations may have been larger than now – European salmon may have swarmed up the rivers from the sea in numbers as great as those of their Pacific relations that jostle up the west coast rivers of America, so crowded that grizzly bears can paw them out of the swift-flowing shallows of the upper reaches. In the Dordogne valley man may have been able to tail great salmon out of the redds by hand alone.

At least some of the mesolithic men who followed those of the old stone age were spear fishermen. The beautifully made fish spears found in abundance on the site of a mesolithic settlement at Starr Carr in Yorkshire show the importance of fish in the diet of these men living in a district of swamps and lakes. The spear-heads, which were evidently hafted in wooden shafts, were cut from slivers of red-deer antler, pointed, and given a serration of barbs by whittling with flint knives and scrapers. They were doubtless used for many hunting purposes, but their primary use was for catching fish, one of the kinds of prey that must be held as well as struck – a land animal, for instance, is better stabbed with a weapon that can be withdrawn and struck in repeatedly. The evolution of the harpoon from such a spear for fish is but a natural step. If the prey is so large that it cannot be lifted from the water at once, the hunter will try holding it with a line, perhaps a thong of rawhide, lashed to the shaft. When the hunter found that that shaft broke or the spear head came away from the shaft, he fixed his line direct to the head and let the shaft go – among the Starr Carr finds there is just such a harpoon-head with a hole for a line bored through its base.

Fishing lines, with hooks, probably came before proper fishing nets, for lines can be made without the use of vegetable fibres; ropes and cords made from plaited animal hair, particularly the long hair of horses' tails or the copious skirts of the yak are in use to the present day. Long lines are, however, easily made from the rawhide of animal skins, just as the South African Boers a hundred years ago made the lengthy reims for their ox-teams by cutting a narrow spiral strip from the hide of an eland; an animal hide of moderate size can produce a strip many dozens of feet long. The sinews from the back-muscles of larger mammals, when split and dressed, also make moderately long and pliable strings of great strength.

The caputre of fish with hook and line must have been invented many times, for fish-hooks of some sort are or were used by primitive people in many parts of the world. The gorge-hook, a splinter tied transversely at its middle to the end of the line, possibly came first, because it is simple and easy to make, though it is inefficient in catching small-mouthed fishes. The conventional fish-hook, which can be made from a twig with a large natural thorn, or even from the top joint of the leg from a large insect such as a big beetle or cricket, if it is armed with a suitable spine, must have come later, though it was not until the use of metal was discovered that the now universal type with barb and sneck became possible. Apart from hand-line fishing on a moderate scale in numberless inshore fisheries, and angling with a rod which is mostly a pastime, fishing with line and hooks has been the basis of several large and important commercial fisheries until the present century.

The most widely known of the line fisheries is probably the great cod fishery of the Newfoundland banks which has been carried on for over four hundred years since the early part of the sixteenth century by ships from Spain, Portugal,

France and other European countries, as well as those from United State and Canadian ports. The annual cod season lasted from June to October during which the small boats and dories lowered from a huge fleet of ships took an immense catch, which was salted and dried to make stockfish and bacalau consumed largely in the Latin countries of Europe and America. The supply of cod seemed to be inexhaustible, so that the yield of the fishery was sustainable for several centuries, until steamships using highly efficient modern methods were introduced and are now taking such a heavy toll that the population of cod is declining. As recently as the late 1930s square-rigged sailing vessels could be seen fitting out for the Newfoundland banks at St Malo and other ports of Normandy and Brittany; but they, like the bluenose schooners of Nova Scotia, are now extinct. This fishery used both hand lines carrying one or a few hooks, and long lines or boulters carrying up to several thousand finer short snoods, each with a hook, at short intervals baited with capelin, squid or herring. Boulters, too, have long been used in the northern parts of the North Sea for catching cod and other bottom-living fishes. After the beginning of this century an important long-line fishery for halibut, a large flat-fish reaching over six feet in length and attaining a great weight, was carried on by steam-vessels off the west of Iceland, in Denmark Strait, and off the coasts of Greenland, taking catches greater than the largest sustainable yield.

The tunny fishery of the eastern Atlantic is very different, though also carried on with hook and line. Unlike the fish mentioned above, the tunny and allied fish are pelagic and live near the surface of the open ocean; they are caught by hooks on lines trailed behind vessels under way, much as mackerel are caught on hand-lines trailed behind small boats. The tunny-men, however, not only tow lines astern and over the quarters, but use enormous fishing-rods guyed out on both sides of the ship in order to increase the number of lines in use at the same time. Tunny are large fish from three to six feet or more in length which apparently swim in shoals of fish much of a size. Large numbers of the smaller fish are taken in Biscay and Iberian waters and of the larger ones around the Canary Islands and further south, though large tunny come north in some numbers in late summer and enter the North Sea. The tunny fishery varies in yield from year to year, but it is uncertain whether the 'bad years' are due to over-fishing, though it has been said that the taking of huge shoals of immature tunny by net off the African coast has diminished the supply of larger fish further north.

The fishery for the equivalent fish of the Pacific, the tuna and similar species, is carried on by the fishermen of the west coast of America and of Hawaii with nets floating in the ocean far from land, not by hook and line. Fishing with hook and line, however, whether the modest hand line or a fleet of boulters spliced end to end for seven or eight miles, whether a small local industry giving a subsistence livelihood or a highly organised and capitalised commerical venture, is infinitely less damaging to fish stocks than fishing with nets. It is, too, much less damaging to the fish themselves; every gastronome appreciates the difference in quality between the firm line-caught fish and those that have been subjected to the crush in the cod-end of a trawl, which inevitably causes some degree of bruising.

Nets are filters, but man learnt to take fish with filter devices long before he learnt to braid a net. A wattle fence of upright rods, interwoven with twigs, makes an excellent weir for catching inshore fish on sandy or muddy coasts with a fair range of tide. Such weirs consist of two long side wings set more or less at right angles, with the apex at about the low tide mark. When the tide flows and submerges the weir the fish swim over the top or round the ends, but become trapped when the ebb uncovers the top and dries the ends. At low water one has only to pick up the stranded fish; in more complex forms of weir a compartment of basketwork at the apex of the V contains the fish. Many elaborations of this simple fish trap are found in different parts of the world, especially where the tide-range is small – long fences guide fish into successively smaller enclosures with re-entrant mouths to prevent escape, and the fish concentrated in the final chamber are dipped out with baskets or hand nets. Weirs seldom take more than modest catches and, although they have been used commerically, they more generally supply only a limited local demand so that they are no menace to the stocks of fish. Such 'fixed engines', as they are termed in English law, are now largely obsolete except for some special kinds of basket-work traps used for catching salmon in tidal waters.

When the use of animal or plant fibres twisted together to make cordage was discovered – and this also must have been discovered many times and in many places – the possibility of making nets arose; could the first net-maker have got his inspiration from a spider's web? No matter whether nets or textiles came first, neither can be derived from the other, for they differ fundamentally; textiles are an interlaced web of warp and weft, whereas nets are tied at each mesh with a sheet bend, the knot used in the modern nets of western civilisation. A simple sheet of netting that could be stretched across a river or stream evolved in two directions, as stationary or set nets and as mobile or drag nets. Set nets could take the place of wattle in fish weirs, or be adapted to make fish traps such as the kettle net and the more elaborate fixed nets used for catching salmon in Scotch coastal waters. When made up as bags they could be set on stakes planted on the foreshore, to catch fish swept in by the tide and left dry at the ebb – the variations of the stow-net and hose-net. The trammel, a loose, fine-meshed net sandwiched between two others of coarse mesh through which the captures push the fine net, each thus making a bag in which it is trapped, is of minor commercial importance, as is the cast net – little more than a peasant fisherman's device.

The important development of the set net in relation to wildlife is, however, the drift net used for catching herring, mackerel, pilchards and other shoaling fish that swim near the surface over deep water. In principle the drift net is simple; a wall of netting put across the path the fish are expected to follow. If it intercepts them successfully the fish push their heads through the meshes when they strike the net, and get stuck because their bodies are too thick to follow and the gills prevent them withdrawing. Corks or other floats support the head of the net, and weights at the foot extend it, so that it hangs like a curtain near the surface, its depth below being adjusted by the fishermen with lines attached to buoys. In modern drift-netting for herrings many nets attached end to end make a fleet a mile or more long; they are usually shot at dusk and hauled at dawn. Drift nets thus differ from the more primitive set nets in not being fixed to stakes in the ground but in drifting free, often many miles from the coast, with the fishing boat riding as to a sea-anchor at the leeward end of the fleet.

Drift-netting for herring has been carried on for centuries in the North Sea – the fishery is so ancient that the time of its evolution is unknown – and an immense tonnage of fish, running to thousands of millions of individuals, has been captured annually with no depreciation of stocks until the recent decline of the present century. Herring feed on minute animals of the plankton by filtering them from the sea-water with their gill-rakers, a sort of grating of small spines on the bony arches supporting the gills – the plankton is swallowed and the water passes out over the gills. Immense shoals of herrings assemble over various spawning grounds off the coasts every year, appearing first off the west and then the north of Scotland, and then successively further south into the North Sea to East Anglia. The fishery for them migrated round the coast from one spawning ground to another, but the fish did not – each shoal belonged to its own ground. The fish using the various spawning grounds form clans or races which experts can identify by minute differences in their anatomy such as the numbers of fin-rays or vertebrae. The eggs of the herring, unlike those of most commercial fishes, do not float in the plankton but sink to the bottom, where shoals of cod, haddock and other bottom-feeding fishes assemble to gobble up the masses of spawn. The survivial of larval herring hatched from the eggs varies from year to year; sometimes there are several bad years in succession, so that the fish taken in each annual fishery are not a fresh crop, but a population mainly produced in one year and becoming increasingly aged until the product of the next successful spawning season begins to replace it.

The stocks of herring taken by traditional methods seemed to be inexhaustible even after the introduction of steam drifters with powered winches about the end of the last century. They provided an important food, mainly in the form of salted herrings, in many countries of Europe, especially in the east. In the 1920s, however, some trawl fishermen found that they could adapt their trawl nets, usually used for catching bottom fish, for sweeping through the herring shoals in mid-water. They made large catches, and trawling for herring caught on commerically. In recent years it has become profitable for catching immense quantities of young herrings too small to be taken in drift nets, and too small for human food, but valuable for extracting the oil and for manufacture into fish meal for use in domestic animal feed-stuffs. This immense drain upon the stocks of young herring is believed to be one of the main causes of the decline of the traditional fishery, and its total failure in some recent years.

Commercially, the North Sea herring is practically extinct.

Great catches of herring are also made with another kind of net, the purse seine, used off the west of Scotland, Scandinavia, Iceland and elsewhere. Like the drift net the purse seine is a curtain of netting supported by floats along its head, but it is used very differently. When the fishermen find a shoal of herring they pay out the net in a circular path to surround it, and when it is safely enclosed by bringing the two ends together they haul in the draw-lines at the bottom of the curtain to make a bag or purse, from which there is no escape as it is brought to the surface to remove the fish. The fisherman formerly found the shoals when they were near the surface by watching for the flocks of gulls and other sea-birds that gather above them to feed. When the shoals swam deeper the fishermen could only shoot their nets at random, though some skippers were skilful at finding a shoal by feeling the fish striking against a lowered lead-line. Now, however, they have a new weapon, for the development of echo-sounding into elegant sonar gear enables them to find the shoals far below the surface, and much too deep to attract the attention of sea birds. Similar sonar apparatus is used also by the mid-water trawlers for finding the shoals.

The herring is one of the most important food fishes by reason of the vast quantities taken in the past, and still perhaps to be taken in the future. But stocks are not inexhaustible; in some places the fishery has passed its peak through taking more than the greatest sustainable yield, so that stocks have dwindled, parts of the industry are ruined, and a valuable food resource has been lost to man. All was well for many hundreds of years until the invention of highly efficient modern methods of finding and catching fish, and particularly of using fish as the raw material for manufactures as well as an article of human diet. These innovations, combined with the increasing demand for all natural products by a rapidly expanding world population, are leading to over-fishing. This will inevitably lead to disaster unless science can discover the size of the crop which can be taken without damage to the stocks, and those interested can agree to respect its limits.

The problems besetting the other fisheries for pelagic fish are generally less than those facing the herring industry. The drift-net fishery for mackerel is much smaller than that for herring, but mackerel are also taken by trawling during the parts of the year when they live in dense shoals near the bottom; in addition line-caught mackerel are sold in the markets of some coastal towns. In summer, shoals of mackerel sometimes come into the bays and harbours of our southern coasts in such numbers that they can be scooped out in baskets and hand nets. The mackerel is a very palatable and nutritious fish, but is notoriously liable to go off very quickly. The English legislators of the eighteenth century, aware of the delicate nature of the fish, made laws permitting the sale of mackerel on Sunday because it would be spoiled if kept until Monday. In recent years the marine biologists have discovered that mackerel are attacked by a disease which breaks down the muscular tissues and reduces them to a soft mush; when a high proportion of fish happen to be infected the loss is serious. The occurrence of this disease no doubt gave rise to the belief that the fish do not keep well; healthy mackerel, though best eaten straight from the sea, last as long as other fish after being caught – drifters and trawlers ice their catches when fishing on grounds distant from their ports of landing. The chief English port for the mackerel drift-net fishery is Newlyn, where great quantities have been landed in past years.

The inshore fishery for pilchards off the Cornish coast has waxed and waned dramatically during the years and, though it was of great local importance from time to time, it was a comparatively small industry. When immense shoals approached the coast they were caught in a net resembling a purse seine, which was drawn round them by a number of cooperating fishing boats. Some of the fish were eaten locally, but the greatest part of the catch was salted, drained of much of its oil, and exported to southern Europe. Sardines, which are merely small pilchards, are taken in vast quantities in fine-meshed drift nets off the Atlantic coast from Brittany to Gibraltar; the fishermen attract the shoals towards their nets by throwing out a bait of mashed fish roe. In waters north of the habitat of pilchards the canning factories substitute small herrings or sprats, but as in England it is illegal to sell other fish as sardines, the canners label the fish 'sild', the Norwegian name for herring, or 'brisling' the Norwegian name for sprats.

The second line in the evolution of nets from the primitive simple sheet of netting led to mobile nets dragged through the water. The simplest form is the seine net, worked from the shore. In using the seine one end of a long web of net is held ashore, and the bulk of the net is payed out from a boat carrying it in a wide sweep or semi-circle until the other end is brought ashore, when the two ends are hauled upon and any fish enclosed are brought in by the centre. The head of the net is buoyed up by floats and the bottom sunk by weights – originally bits of wood and stones. It would not have been long before fishermen using this kind of net found the advantage of making the centre of the net in the form of a bag to increase their chances of taking a good draft of fishes. Seine fishing is an ancient way of netting fish; seines have been used for thousands of years, and in various forms are found in many parts of the world.

It was probably late in the eighteenth century that fishermen in the English Channel invented the trawl, by towing behind a boat the equivalent of the bag part of the seine, with its mouth held open by a beam. As the trawl was improved, the immense advantage of fishing at sea, farther out than the usual few hundred yards along the shore, was obvious when the teeming offshore grounds began to be harvested. By the middle of the nineteenth century the trawl net towed behind a decked fishing smack was in wide use. The net is an enormous conical bag tapering to a point, the cod end, where the fish accumulate and which is secured with a lashing that is undone to release the catch when the net is hauled on deck. The mouth of the net is kept open by a wooden beam or pole, up to forty feet long, socketed into iron frames or shoes at each end so that it is held several feet above the ground. The head of the net is laced to the beam but the foot rope, much longer than the beam and attached to the bottom corners of the shoes, trails back in a deep U-shaped loop. When the net is towed the upper part of the net fixed to the beam overtakes the fish on or near the bottom, and forms a sort of roof above them, before they are disturbed by the foot rope over which they are swept towards the cod end. Various valves of netting inside prevent the fish escaping, and old netting, canvas, or cow hides portect the underside of the cod end from chafing on the bottom.

Beam trawling almost certainly first began in the ports of south Devon, especially Brixham, from which it spread both east and west; but it was not until the 1840s and 50s that the west country men came up Channel and began trawling in the North Sea. At first they used the east coast ports such as Lowestoft, Yarmouth and Grimsby only as seasonal bases, but many of them migrated permanently in later years, bringing their families to new homes in the east. An immense trawl fishery exploiting the richly stocked grounds of the North Sea, with sailing smacks using beam trawls, grew up in the second half of the nineteenth century after the railways came to the east coast ports, so that landings were quickly delivered to London, the towns of the Midlands and of the industrial north. On the grounds far from land, as on the Dogger Bank, large fleets of trawlers fished together under the command of 'admirals', and daily ferried their catches in their row-boats to swift sailing cutters, which rushed the fish, packed in ice, to Billingsgate market where it arrived in finest condition. Railways, factory-made ice, and the fast cutters, beside the enterprise of the fishermen and smack-owners – often the same men – and the vast quantities of fish for the taking, were the foundation of the trawling industry, which produced huge amounts of good food for a rapidly growing population.

Sometime about the 1870s another form of trawl was invented, the otter trawl, but it did not come into commercial use until the last decade of the century. The net is similar to that of the beam trawl; it is extended not by a beam, but by a large board fixed to each wing and arranged so that when towed the pressure of the water makes the two boards fly apart like kites, thus keeping the mouth of the net open. It was adopted universally by steam trawlers soon after they were introduced in the 1890s because, not being limited in width by the maximum practicable length of beam, it could be made very much larger, with a mouth a hundred feet wide or more. The steam trawling industry grew rapidly, taking the place of the sailing beam-trawlers, a few of which nevertheless lingered on until the outbreak of the Second World War. The steam trawlers, with the enormously increased efficiency of their gear and much greater radius of action, at first exploited the fishing grounds of the North Sea and the English and Irish Channels. They soon went further afield to the prolific banks of the Celtic Sea south of Ireland, those off the west coasts of Ireland and Scotland, and the Rockall and other banks far out in the Atlantic. Larger and more powerful trawlers began voyages to the fishing grounds of the Faroes, Iceland, Bear Island, the Murmansk coast and the entrance of the White Sea, and some went as far south as the coasts of Morocco.

At the same time various developments of the seine, sometimes as a sort of hybrid with the otter trawl, as in the Danish seine towed by a ship and not used from the shore, or the net towed by Spanish fishing vessels working in pairs on the Great Sole Bank south of Ireland and elsewhere, became the instruments of important fisheries. The intensity of fishing led to a marked decline in the harvest of the sea by the beginning of the First World War but, though stocks of fish increased during the war-years when fishing was restricted, they fell off again after a few fat years when fishing was resumed. Yet in spite of the immense quantities of fish landed over the years, and the consequent depletion of stocks, there are still enough fish in the sea to support a huge industry. Some effort is being made to learn from past experience to avoid over-fishing, so that the industry shall not ruin itself, and to ensure that a regular source of food can be maintained. The extension of territorial limits by most nations in recent years brings some control over the exploitation of fishing grounds that were once free for all, and numerous international agreements have been made in the hope of preventing over-fishing. Since the early years of the twentieth century the International Council for the Exploration of the Sea has helped in coordinating the activities of the government and other marine laboratories of European countries on all aspects of ocean research, with fruitful results for the exploitation and conservation of the fisheries.

Fishermen who follow a hard and hazardous calling wresting a livelihood from the sea, at first paid little attention to the results of scientific research – their lack of education and their way of life gave them little opportunity, or inclination, to understand and heed what knowledge of the biology of the sea could offer. Since the end of the Second World War, on the other hand, the application of science has made great advances in the fishing industry, although much of the innovation has been directed at increasing the yield of the fisheries rather than their conservation. Larger ships, using greatly improved gear for catching and deep-freezing the fish, search for their prey with sonar and echo-sounding electronic apparatus and use radar navigation aids. The men themselves are, more and more becoming highly skilled technicians, enjoying amenities when afloat undreamed of by their ranting and roaring, brute force and ignorance, predecessors. Some of the larger vessels are factory ships in which the fish is prepared on board for the frozen food market, though much of this work is also done at the ports where the catches are landed.

The modern fishing industry is carried on by the ships of many nations not only in the eastern north Atlantic, but everywhere in the world that can produce useful catches. It is generally most vigorously prosecuted near the large centres of population, though deep-freeze and other factory vessels can operate profitably in more remote regions. Both coasts of the Americas, the southern ones of Africa, the Indian Ocean and the north Pacific, and even the Antarctic make enormous contributions to the sea-food supplies of the world.

The great distance from home to which some nations send their fishing vessels is surprising; the free-port of Las Palmas in the Canary Islands is used as a fishing base by the ships of many nations, and the harbour is crowded with ships from Spain and other Mediterranean countries, and with large stern-trawlers from Russia, Japan and Korea. It is widely felt that an indiscriminate struggle between the nations, and between rivals within the nations, will certainly lead to over-exploitation of the fisheries, and to reduction of the stocks of fish below the sustainable yield. The work of some committees of the International Biological Programme, which has been in progress since 1965, is aimed at examining some of the problems of organic production in freshwaters and the seas, and of the potentialities and use of new as well as existing resources. The results will, it is hoped, pave the way for further research, and help to show where exploitation can be combined with conservation of stocks.

Although the fishing industry has in the past often been conservative and reluctant to adopt new methods or to seek new quarry, it has needed no advice from outsiders to make it respond to the demands of the market for more and cheaper fish. In the early years of the twentieth century the market for many kinds of fish was limited by prejudice, particularly in Great Britain, and much potential good food was thrown away by the fishermen who knew that it would not pay them to bring it ashore. The increasing prosperity of the masses has now widened the demand, so that many kinds of inferior fish are now acceptable. Probably the first despised species to be widely sold was the dogfish, which when beheaded, skinned and disguised under the names of 'rock salmon', 'rock eel' or 'flake', is consumed in large quantities by the fried fish trade.

The coalfish or saithe, a fish of little culinary merit, was formerly rejected by British fishermen though landed at Continental ports – I have seen literally tons of them thrown over the side on the Iceland grounds – but is now readily sold as 'coley'. It is a relative of the cod, and its silvery sides and compact shape when first taken from the sea give it a look of quality belied by its cotton-wool-and-fish-juice flesh when cooked. The bright red Norway haddock or 'soldier' is another species formerly rejected by the British market, but accepted elsewhere; attempts to popularise it with the British housewife under the name of 'red fish' have been only moderately successful.

One crustacean – shellfish in fish-trade language – that was formerly little regarded has become popular since the Second World War; the tails of the Norway lobster or Dublin Bay prawn are esteemed as 'scampi', though they are not in fact the same species as the Mediterranean prawn to which the name rightly belongs. In the early 1920s I saw enormous quantities brought up in the trawl from a hundred fathoms south of the Westmann Islands off Iceland; the fishermen threw them away as unmarketable – they did not even know they were edible. I got the cook to boil some, and the crew were surprised at how good they were; thereafter we had scampi at every meal, and the cook pickled a bushel of them in vinegar for the voyage home.

Among the larger crustacea, crabs and lobsters have always been popular, but although great numbers are taken – generally in 'pots' traditionally made of wicker-work, but now largely superseded by less clumsy traps – the fishery is an inshore one and small in comparison with the deep-water trawling industry. In recent years skindivers with Scuba gear have taken to catching lobsters and crabs on rocky coasts for a livelihood. It is peculiar that the crayfish or spiny lobster, the *langouste* of the French, has never been popular in England although it is common on the south-west coasts so that the 'Johnny crabbers' from Brittany used to exchange their lobsters for crayfish with the Cornishmen. In other parts of the world crayfish tails are the subject of a large canning industry, and factories have even been started in such remote islands as Tristan da Cunha in the south Atlantic and Amsterdam Island in the south Pacific. In the north Pacific there is a large Japanese fishery for the giant spider-crab, which reaches a size of over six feet across the outstretched legs; the crab's meat is canned, but the labels on the cans show pictures of the familiar edible crab – a portrait of the spider-crab would not be likely to attract buyers.

Shrimps and prawns of many species, among the smaller crustacea, are valued as food; most of them live on or near the bottom in shallow water and consequently their capture is an inshore fishery. They are usually taken with small trawls worked from boats within a few miles of the coast, but are also taken in tidal stake-nets, and formerly in smaller hand-nets pushed by men or towed by horses wading near the beach. Nowadays shrimp nets with kite-like boards, to keep them running parallel with the shore, are towed by tractors on the beach in places with wide shallow sands such as Morecambe Bay. A few deep-water kinds of crustacean are marketed, such as the bright scarlet prawns with large eyes from the Norwegian fjords, which look better than they taste; but of them all the shallow-water shrimp from the Baltic, called 'reie' in Danish and Norwegian, is probably the most exquisite in flavour.

Although the shrimp and prawn fisheries are not as capital-intensive as the deep water fisheries the total world catch is immense. The planktonic shrimp-like euphausian or 'krill' of the Antarctic, which is the main food of the whalebone whales, is now being directly exploited. Now that whalers have over-fished the whales to the point of near extermination of some kinds, and of disastrous reduction of the stocks of others, Russian ships are taking huge quantities of krill by pumping the teeming shoals on board for manufacturing into food for man or animals – they no longer wait for the whales to process the krill into commerically valuable products. Potted krill is the new rival to caviare.

Fisheries for the true shell-fish, the molluscs, are also generally inshore fisheries of local importance, but in the aggregate their production throughout the world is great. Many kinds of bivalve molluscs are greatly valued as food or luxuries, from the large scallop of Europe – the coquille St Jacques, the badge of pilgrims returning from the shrine of St Iago de Compostella; the cockles and mussels; the palourdes of France; the clams of America and the gooeyducks of the north-west Pacific coast; to the universally esteemed oyster, of which one kind or another is found in all but the colder regions of the world. As well as the harvest of shell-fish from the wild, artificial cultivation, particularly of the oyster, gives a crop larger than any provided by unaided nature.

Oyster culture is carried on in tidal and shallow seas, frequently on the shores of large estuaries, by providing good settling-places for the planktonic oyster-larvae to come to rest when they become sedentary as young oysters. The cultivators then plant them out in oyster parks, where they are not overcrowded but have ample room to grow on the rich plankton which they filter from the sea water. When ready they are put out on the fattening beds for harvesting at the age of three or four years. The industry is said to have started in the land-locked bay of Arcachon in the south of the Bay of Biscay when a ship carrying a cargo of oysters from Portugal was wrecked there in the mid-nineteenth century. Oyster culture is both a science and an art. and has evolved into a highly skilled and profitable industry supplying immense quantities of valuable food to the markets. The culture of the English native oyster is pursued in a somewhat similar way; young oysters are planted out on beds, generally in estuaries where the water, fertilised by the drainage from the land, supports a rich growth of microscopic plankton. Cultivated oysters, like plant-crops on the land, are attacked by many pests, and the cultivators have to carry on an unending fight against starfish, the whelk-tingle or sting-winkle, the oyster-drill, and the slipper-limpet, the last two accidentally introduced from America to the English oyster beds, all of which feed on oysters and are most destructive to the fishery.

There are, however, molluscs that are not shell-fish, the octopuses, cuttles and squids, which are some of the quickest-growing and most plentiful animals in the sea. They are caught with hook and line, nets, and in simple traps. The octopus trap used in the Mediterranean is merely a globular clay pot with a wide mouth; the fisherman lowers to the bottom a long line with pots attached at intervals, and draws it up after sufficient time has passed for octopuses to take up residence in the pots, from which they do not try to escape when raised. Octopuses live on the bottom, cuttles near it, but squids live in mid-water and are taken mostly by nets. These creatures are highly esteemed as food in many parts of the world, but in Great Britain they are little regarded, though in demand by the long-line fishermen as bait for their hooks.

Some shell-fish are taken for uses other than food – the extensive fishery for pearl-oysters in the Pacific and Indian Oceans and in the Red Sea harvests a large tonnage yearly. The oyster is sought not only for the pearls, obtained by letting the bodies of the animals rot in the sun, but for the shells, which consist largely of mother-of-pearl used for many decorative purposes and particularly for making buttons. The shell of the chank, a large snail-like mollusc of the Indian coasts, is similarly used. As is well known, Japanese biologists have devised ways of stimulating the development of pearls in cultured pearl-oysters to obtain large quantities of real pearls. Some people allege that they can distinguish cultured from natural pearls, but as both kinds are formed by the same biological process the claims of such self-styled experts may be doubted.

The artificial culture of shellfish in shallow waters depends for its success on the sedentary nature of the animals, which cannot escape; but the culture of free swimming fish, which can, presents many problems. Experiments have shown that young fish can be reared to marketable size in enclosed lagoons, and that they grow more quickly than fish at large, which have to find their food and escape from predators. Trials have also been made on releasing into the sea young fish that have been artificially reared beyond the very vulnerable early stages, but no large commercial development of cultivated sea-fish has yet been successfully accomplished. In a few places the warm water discharged from power stations after being used to cool the condensers has been used for quickly rearing fish fry or shellfish, and for cultivating shrimps.

With freshwater fish, on the other hand, there are fewer difficulties, so that fish-culture has been carried on in many parts of the world, particularly in the East where it has been practised for several millennia. The luxury market for trout in Europe is supplied with fish reared in captivity, where they are fed largely on slaughter-house refuse. Many species including vegetarian ones such as carp, are or have been artificially cultured in the East, in natural ponds and man-made tanks. Some of the numerous species of *Tilapia* or lake fish grow fast when fed with cut grass thrown into the rearing ponds, but they tend to breed before they have reached their full potential size so that overpopulation produces great numbers of small fish, whereas an equal weight of a smaller number of large ones is preferred. The desired result has been attained through the discovery that the hybrid offspring from a cross between two of the species are all males. The fish in a pond stocked with the hybrid therefore cannot breed; their numbers cannot increase, they do not compete for food and consequently reach a large size in a short time.

The drainage from the land, too, provides nutrients for the planktonic and other plants that form the start of the food chain in freshwaters. The run-off from the land, however, is sometimes so heavily contaminated with excess organic fertiliser, intended to stimulate a high yield from cereal and other crops, that the waters are over-fertilised – eutrophicated, in technical jargon – so that vast quantities of various algae fill the water to the detriment or even destruction of the fauna.

Although the waters of the oceans have not been deliberately fertilised artificially by man, one may wonder whether the productive fishing grounds of the North Sea, into which the sewage of half Europe has been poured for centuries, may be the result of unintentional fertilisation. Now that industrial waste has been added to the sewage, pollution with various substances such as heavy metals and organic poisons can be recognised far and wide, from which we may infer that the harmless nutrients are equally widespread and may help the seasonal growth of plankton. Some have said that the dilution of sewage by the waters of the North Sea is so great that any fertilising effect is negligible. Yet the building of the Aswan High Dam in Egypt has brought widespread ecological damage, not only to the land, but also to the Mediterranean sea, where the sardine fishery of 18,000 tons annually has come to an end because the annual flood of nutrients brought down from Nubia has stopped. The nutrients, too, no longer fertilise the farms of the Nile valley so that the peasants must buy artificial fertilisers if they can find the money, whereupon the waters become contaminated and suffer from eutrophication.

Many smaller industries, as well as the great commercial fisheries, take their toll of the wildlife in the sea. In some of the warmer seas fishing for sponges is extensive, though of less importance since the introduction of synthetic substitutes. In the Pacific there is a flourishing fishery for the large black sea-cucumber, a relation of the starfishes and sea-urchins, which when sun-dried is used in great quantities in the cookery of the Far East under the names of 'trepang' and 'bêche de mer'. The eggs of all the seven species of sea turtle are taken in such great quantities from the sandy breeding beaches of the tropics, where they are laid in nests dug out by the mothers, that it is a wonder that turtles have not been exterminated. They are, indeed, threatened, but the authorities of the countries where turtles breed are at least beginning to make protective laws which, if enforced, will prevent the gathering of a crop in excess of the sustainable yield. Experiments have shown, too, that turtle farming may become a profitable industry. The eggs of some kinds of turtle are esteemed as a culinary delicacy, but the main use for them is the extraction of oil. Similarly the eggs of freswater tortoises, often wrongly called turtles, are gathered in millions from the sandbanks where they are laid, particularly in the great rivers of South America.

In addition, the flesh of at least one kind of turtle – the green turtle – is much sought after as a luxury food, and with the advent of the canning industry the exploitation of the species has increased, greatly to the detriment of the numbers of its world population. The flesh of the other turtles is less esteemed, but the horny plates on the shell of the hawksbill turtle are the tortoise-shell of commerce, so highly prized that had synthetic substitutes not been invented the species would be nearer to extinction than it is already.

The cult of shell-collecting, once the privilege of rich eccentrics, is now widespread. It has grown particularly with the rise of the package-tourist industry, in which dealers sell shells as souvenirs to tourists ignorant of, and indifferent to, the devastation they are bringing upon the wildlife of the world. It has even been suggested that over-fishing the beautiful *Strombus* to sell its shell to tourists has been responsible for the spectacular population explosion of the Crown-of-Thorns starfish in the Pacific and Indian Oceans and the Red Sea. The *Strombus* is the main predator on the starfish, which feeds upon the polyps of coral, and which during recent years has increased in numbers to plague proportions so that great areas of coral reef have been destroyed. On the other hand it is possible that the outbursts of hordes of the starfish may be due to natural causes beyond the influence of man.

Tropical coral reefs have recently been seriously damaged by the increase in dynamite fishing off the shores of the developing countries. It is so much easier and quicker for lazy people to throw a few sticks of dynamite overboard and then collect the dead and stunned fish that float to the surface after the explosion, than to undergo the labour of hauling nets or baiting and setting fishing lines. In a few moments large parts of the submarine environment are destroyed, leaving a devastation which cannot be repaired for many years and may never be restored to its original state.

Dynamiting at least has the excuse that it produces food, but the same cannot be said for the sport of underwater pleasure-hunting by Scuba divers. It is unfortunate that no sooner has man discovered a way to visit a hitherto unknown region, of the greatest natural beauty and interest, than he has immediately to start destroying it. It is no more than vandalism to spear some huge and ancient fish, that has lived in its shelter among the rocks for perhaps half a century, and drag it ashore to be photographed and weighed in the hope that it may be a 'record'.

A recent enquiry by the Natural Environment Research Council into the need for marine wildlife conservation found that the activities of skin divers are the commonest cause of the degradation of the sublittoral fauna. The destruction is brought about not only by diving for sport, but also by diving for the commercial gathering of shellfish, mainly lobsters, and for the collection of inedible animals whose dried remains are sold to tourists. On the other hand the Council also found that change in the climate is the next commonest known cause of downward changes in the abundance of the fauna. Although much has been said about marine pollution as a cause of damage to the flora and fauna, the Council found that pollution was reported less often as adversely affecting the fauna. Several countries have reserved sublittoral areas of various sizes for complete protection, or controlled exploitation, with results that are reported to be highly successful; interest in the subject is spreading, so perhaps further protection may be given before damage becomes irreparable.

The seas are so vast in three dimensions that it has seemed that their traditional harvest is inexhaustible – although it is not a true harvest, for here man reaps where he has not sown. All was well when the human world population was moderate in size, and while fishing methods differing little in principle from those of stone-age man were used. Now that the human population is said to be nearing disaster proportions, the exploitation of the seas with new and better ways of catching sea-food animals in huge quantities is beginning to take more than the sustainable yield of many species. The obvious remedy is to refrain from overfishing by international agreement. Apart, however, from the fact that international agreements are not easily made or kept, because the nations are rivals contending for the resources of the oceans in a kind of international rat-race, we do not have the knowledge of the maximum sustainable yield for most of the products of the sea. Applied marine biology has an enormous task ahead in discovering the population dynamics of the commercially desirable sea animals, so that rational exploitation may follow. In the meantime such agreements as are reached must be based empirically on incomplete information – they may be better than nothing, but they can be worse than anything if they bring a false feeling that all is well, and that nothing more need be done.

4 THE ATTACK ON THE MARINE MAMMALS

In addition to the 30,000 different kinds of fishes and the hundreds of thousands of invertebrate species down to the protozoa, a few kinds of higher vertebrates, little more than a hundred in all, provide or provided an economic resource much greater in value than their small number might imply. The thirty-one species of seals, eighty-eight of whales and dolphins and porpoises, and four of sea-cows are warm-blooded, air-breathing mammals, descended from ancestors that lived on land but took to an aquatic way of life to which they have become highly adapted. They have given a harvest of meat, oil, leather, and many other useful things and, because most of them are gregarious at some time of their annual life cycle, are often available for exploitation in large numbers. The seals come ashore or climb on to floating ice to breed, and consequently can be attacked on foot, but the whales are completely aquatic and must be pursued in ships and boats.

The upper palaeolithic men of the Dordogne were familiar with seals, as is shown by the drawings they left. One engraving in particular is so true to nature that the species shown can be readily recognised – a bull and cow grey seal. One may infer from these pictures either that the men of the rock-shelters were in the habit of travelling to the sea-coast, or that grey seals came far up the rivers – which were probably much wider and deeper than now – perhaps in pursuit of salmon, or that engraved artifacts travelled far by passing from hand to hand possibly in barter. Nevertheless bones of seals are not plentiful in palaeolithic occupation sites nor, more surprisingly, in shell middens, although seals were probably at least as plentiful along the shores of Europe in those times as they are today. The occasional caputre of a seal was no doubt put to good use, but it seems likely that easier ways of earning a living than hunting seals were preferred. In the harsher conditions of the Arctic, where seals are abundant, the Eskimos, who were living in a stone-age until a few centuries ago, made seals and their products one of the mainstays of their economy. But the number of Eskimos was comparatively small while that of the seals was great, so that hunting had no adverse effect on the stocks of prey.

Similarly, until recently on the western coasts of Scotland and Ireland crofters and subsistence-farming and fishing peasants hunted seals for their oil, hides, and other products, generally by making annual raids upon the breeding colonies, but on a scale too small to exterminate the stocks. This hunting was not done commercially to produce things for sale but to provide articles for domestic use. The harvesting did, nevertheless, affect the size of the population of seals, which remained fairly constant and below the potential level that the environment could support. The end of subsistence-farming and fishing with the coming of the Welfare State, which offers a soft living in place of hard work, released the small hunting-pressure on the seal population, which in a few years exploded to become a menace to commercial fisheries. While the seals, especially the grey seals, were increasing to such numbers that fishermen came to look upon them as pests, the British Parliament passed an act prohibiting the killing of seals – to the disgust of fishermen but the delight of well-meaning though ignorant people most of whom had never seen a seal in the wild.

Both the species of seal found commonly on the coasts of Great Britain, the grey seal and the smaller common seal, are becoming increasingly serious causes of damage to fisheries. The weight of commercially important fishes, in addition to other kinds of fish and shellfish, eaten every year by the population of seals numbering many thousands runs to hundreds of tons, though it is doubtful whether this toll seriously affects the amounts caught by fishermen. Grey seals are certainly a nuisance to the men who catch salmon in nets fixed close to the shore along the east coast of Scotland, for they eat or mutilate many of these valuable fish – they formerly damaged the nets extensively, but since the introduction of artificial fibre, which is much stronger than hemp or cotton, for net-making, the damage has decreased.

Far more serious than preying upon fishermen's potential catches and damaging nets, is the indirect damage inflicted upon 'white fish', particularly the cod. Seals carry great numbers of parasitic worms in the intestine, round thread-like worms of the class Nematoda, whose eggs are passed out into the sea water in the faeces of the animals. The larvae that hatch from the eggs do not directly infect seals, but get into various fishes, especially the cod, where they pass into the flesh and coil up among the muscle fibres in a dormant state. There they remain, and cannot complete their development to adult worms unless the cod is eaten by a seal, whereupon digestion releases them from the fish's flesh and they become full-grown breeding worms. The proportion of fish infected with the cod-worm has increased enormously in recent years in parallel with the expansion of the seal population, much to the detriment of the fish trade. Most of the cod sold in fishmongers' shops are infected with cod-worm, and though the probability of finding worms in the flesh has been hushed up as much as possible, lest the housewife should give up buying cod, consumers are beginning to know what to expect. The worms are in no way harmful to a person who eats them with his cod, but they are aesthetically so unpleasing as to fill some people with disgust when found on their plates. The firms that prepare frozen cod-fillets for packaging and sale from deep-freeze cabinets go to great trouble and expense to examine the fish fillets and remove the cod worms, for they fear that the discovery of worms in their products would damage their business.

Some seals are thus now becoming vermin because they damage man's interests by direct and indirect effects upon the fisheries; but for the rest man's relationship to seals has nearly always been the reverse – he has been the aggressor and the seals the victims. Small-scale subsistence hunting, as already pointed out, did not threaten the stocks of seals, but when civilised man explored the remoter regions of the world and found huge populations of seals that had never been disturbed, commercial sealing began and quickly brought some species to the verge of extinction. Seals were hunted for the oil in the layer of blubber that encases their bodies beneath the skin, for the leather that can be made from the hide, and for the fur in which some species are clad.

The seals form two natural groups, the 'true' or earless seals that when on land creep on their bellies trailing their hind limbs, and the 'eared' seals or otaries that can turn the hind limbs forward under the belly and raise the body from the ground, so that they have considerable agility when on land. The two kinds differ in the ways they swim; the true seals mainly use side-to-side strokes of the hind flippers, whereas the otaries mainly use rowing strokes of the comparatively long and powerful front flippers. Most species of both groups are gregarious in the breeding season, and haul out in great numbers on the sea shore or on floating ice for the females to give birth, and to be inseminated by the males to conceive the next year's crop of young – generally termed 'pups' though their parents are termed 'bulls' and 'cows'.

The walrus, which is classified in a separate suborder because it does not exactly meet the definition of either of the two main groups, was one of the first species to be commercially exploited. It has always been a prized prey of primitive peoples inhabiting the Arctic, but in 1610 hunting walruses for their oil and ivory was one of the objects of the Muscovy Company's expedition which discovered Spitzbergen, where the unexpected find of abundant right whales started the Arctic whaling industry. In the centuries since then the walruses of both the Atlantic and the Pacific Arctic, which differ slightly, have been greatly reduced in numbers, chiefly by whalers who combined whaling and walrus-hunting. Indeed, the American whaling fleet killed so many Pacific walruses that the Eskimos, to whom the walrus was a mainstay of life, were deprived of their living so that many died of starvation. The immensely thick hide on the neck of the walrus was valued commercially, as well as the ivory tusks and the blubber-oil; until recent years it was in demand for making buffing polishers used in the metal industries.

The harp seal, which annually hauled out in enormous herds on the floating ice of the north Atlantic to give birth to the pups, was one of the first species of true seal to suffer commercial exploitation. Harp seals are still hunted on their breeding places in the European Arctic by the Norwegians and Russians, and in the west by the Newfoundlanders and men from the Canadian maritime provinces. The new-born young are the main object of pursuit though a lesser number of adults is included in the catch. The fishery has been carried on at least since the early part of the sixteenth century, and reached its peak in the second half of the nineteenth, when annual catches of from 800,000 to 900,000 seals were made. The seals have not been exterminated by this enormous drain on the population, and the fishery continues to the present day, though catches are much smaller. Some protection is now given to the herds by regulating the dates of the fishery and the numbers to be taken. Objections to the mass slaughter have been raised on humanitarian grounds, because the wholesale spilling of blood distresses some people. Those engaged in the work, however, do not appear to behave with wanton cruelty, though such a trade inevitably breeds a measure of callousness in those practising it.

There are no otariid seals in the North Atlantic, so it was not until the coasts of the lands in the southern hemisphere and the north Pacific were explored that the herds of otaries living on those shores could be disturbed by Europeans. The different kinds of otary fall into two groups, the fur seals and the hair seals or sea-lions. The first bear a dense undercoat of soft hard-wearing fur of high commercial value; the coats of the second consist of harsh stiff hairs, so that the hide and the thick blubber under the skin are the commercially valuable products. In many of their haunts both kinds are found together on the beaches where they assemble in great numbers to give birth to their young, the cows gathering in groups or 'harems', each the personal property of a bull which drives would-be rivals away. It is the polygamous habit in both eared and true seals that gives man the opportunity to manage and conserve the stocks.

Fur seals and sea lions are found on the southern coasts of South America, Africa and Australia and on the shores of numerous subantartic islands round the globe. They also occur at many places on the Pacific coasts of the Americas and the offlying islands, from Juan Fernandez through the Californian Islands to the Pribilof Islands in the Bering Sea. From the Pribilofs the distribution extends on to the north-east coasts of Asia. Soon after Bering's voyage of discovery in 1741 Russian hunters and traders were exploiting the millions of fur seals that came annually to the Pribilof group for breeding. During the last half of the eighteenth century and the first half of the nineteenth enormous numbers of fur seals were killed, and at various times the population was so reduced that the Russian government restricted, and sometimes suspended for a year or two, the wholesale killing. One of the reasons for the later intense exploitation of the fur seals on the Pribilofs was the near extermination of the fur seals of the southern hemisphere in the first quarter of the nineteenth century, as related below. After Russia sold Alaska and the islands to the United States in 1867 the killing was done in a more sensible way, so that a large yield was taken without damage to the stock, which consequently grew again to an enormous size numbered in millions.

As the otaries are polygamous, single bulls possess many cows and father many pups; the pups of each sex, however, are born in approximately equal numbers, so that there is an excess of males. It is the excess of males, killed when they are about three years old, that makes the annual crop, which should continue to be available indefinitely, barring some unforeseen ecological disaster. This satisfactory state of affairs was before long threatened by the efforts of pelagic sealers from several countries, who hunted fur seals on the high seas where they were subject to no control. Their wasteful methods of hunting destroyed large numbers of pregnant cow-seals, and failed to retrieve many of the animals killed in the water. Not only was the United States monopoly threatened, but the very existence of the seal herds. The problem posed by pelagic sealing was finally resolved by the signing in 1911 of an international treaty between the United States, England, Russia and Japan, in which all the signatories undertook to require their subjects to refrain from hunting northern fur-seals on the high seas. As the United States is the sole beneficiary of the agreement the other signatories are paid compensation calculated as a proportion of the annual revenue accruing to the United States government from the seal-skin industry. The proportion paid to each is reckoned in relation to the value of their former pelagic catches. The yearly payments continue to the present day.

The mad exploitation of seals in the southern seas was far different from the rational management of the industry that evolved under the United States government in the Pribilofs. The explorers and world-voyagers of the seventeenth century found seals plentiful on many of the southern coasts, but it was not until after Captain Cook's circumnavigations at the beginning of the last quarter of the eighteenth century that commercial sealing grew into a large industry. Some sealing had been done long before in central and south America; towards the end of the seventeenth century, when the Caribbean area was being ruthlessly exploited by the Spanish colonisers and pirates, buccaneers, log-wood gatherers, and others of various nations, the monk seals that bred on some of the sandy cays were killed off and boiled down for oil, which was traded locally rather than exported to Europe. Monk seals are not otaries but true seals which live in warm and tropical seas. The three species are now scarce, probably because they were much more accessible to man than the seals of the far-off southern ocean. The monk seal of the Mediterranean still exists in the Adriatic and elsewhere, and on the Atlantic coast of Mauritania; the Hawaiian monk seal of the Pacific is reduced to a population of a few hundred; but the Caribbean monk seal was so heavily attacked that it was for long thought to be extinct, until a few lone survivors were seen in recent years.

Captain Cook found great herds of fur seals on the shores of such subantarctic islands as South Georgia in the south Atlantic, and within a few years of his return to Europe when his discoveries became known many sealing vessels were following his tracks to gather the rich harvest that was there for the taking. The rise of the sealing industry was like a gold-rush, with everyone ruthlessly grabbing all he could and the devil take the hindmost. Such vast numbers of seals were killed for their skins that the teeming herds were soon so reduced that the sealers were ever searching for new coasts and islands to ply their trade. The fur seals on the southern coasts of South America, South Africa and Australia, all species closely resembling each other, those of Kerguelen, South Georgia, the subantarctic islands of New Zealand, and many other islands were reduced so near to extinction that the sealers took to hunting the huge elephant seals, which are true seals, not otaries, that came ashore to breed on the coasts of the sub-antarctic islands. Elephant seals or 'sea elephants' bear no fur but have coarse hair on the skin, beneath which they carry a thick layer of insulating blubber that yields an oil less profitable to sealers than fur seal skins, but of sufficient value to supplement their cargoes when fur seals were scarce. Elephant seals were also hunted by the whalers to augment their catches of whale-oil.

The fur-sealing industry flourished for less than fifty years and destroyed itself through the indiscriminate slaughter of every animal the sealers could find, until so few were left that hunting them was no longer profitable. A single exception set an example that, if it had been heeded, might have perpetuated the industry indefinitely. The sealing rights on the Lobos islands at the mouth of the Rio de la Plata were farmed by the governor of Monte Video and leased to sealers on the condition that they kept a close season and that cow seals and their young were not to be killed – the sealing was managed on lines much the same as those later used in the Pribilofs. It was possible because the islands were within territorial jurisdiction and could thus be policed; the remote islands of the southern oceans and the coasts of the continents, whatever sovereign rights might be claimed over them, were in effect no-man's lands where uncontrolled exploitation by the sealers and whalers could run riot.

Perhaps the most notorious example is provided by the history of sealing in the South Shetland Islands, which were accidentally discovered by the Englishman William Smith in 1819 when bad weather drove him off course far to the south on a voyage round the Horn from Montevideo to Valparaiso. Smith saw great numbers of seals on the islands and reported his discovery in confidence to the English consul at Valparaiso, hoping that the new sealing grounds might be exploited solely by his countrymen. But the secret leaked out, and in the following year a fleet of American as well as English sealers descended upon the islands – the Americans of the New England ports were from early days pioneers in whaling and sealing, and always took the largest part of the industry. In spite of the loss of several ships, both American and English, the seals were killed in such numbers that in only three years the animals were nearly exterminated, and the South Shetland sealing industry came to an end in 1822 after harvesting probably over half a million seal skins.

Sealing continued throughout the nineteenth century and into the twentieth, but it was a poor trade compared with the get-rich-quick prosperity of the early days. Small sailing ships, mostly schooners or brigs, picked up a few skins here and there but relied mainly upon sea-elephant oil for their profits. Life in the ships was extremely rough, and rougher still in the hovels in which sealing gangs existed when they were put ashore on remote islands and left, generally for some months but sometimes for several years, to kill seals while the ship cruised on to hunt elsewhere. The fur seals, sea-lions and sea elephants thus never had the chance to rebuild their numbers, for as soon as a local increase occurred in a place abandoned as unprofitable and left unmolested for a few years, a raider arrived and wiped it out. By the end of the nineteenth century the industry was almost extinct, and was carried on by only a few ships whose masters had been sealing all their lives and knew no other trade.

At the beginning of the twentieth century elephant seals had been for some time so scarce that hunting them was no longer profitable. They had consequently been little disturbed for twenty years, and in that interval they had increased in numbers, particularly at South Georgia. Great Britain asserted her sovereignty over this island when the southern whaling industry started, and allowed the hunting of elephant seals under licence which permitted the killing of adult bulls only. Under a simple system of management the seal population grew rapidly, so that the herds are now as large as they ever were. Lack of interest in 'elephant oiling', combined with control and protective regulations by other governments, led to similar population growth in other islands throughout the southern ocean, now no longer forgotten and neglected by nations claiming them as their territory.

Although the elephant seals came back, the fur seals seemed to have been brought so near to extinction that it was doubtful whether they could recover. A few were seen from time to time by whalers, and small populations were known to exist on some of the islands and the continental coasts of the south, but it was only the South African species that maintained its numbers, because its exploitation was early brought under government management. But although almost unnoticed, the numbers of fur seals were slowly increasing, and when the growth of the population turned the corner of the exponential curve it suddenly soared upwards with an explosion that surprised everyone. On one part of South Georgia a herd of 40,000 fur seals have their breeding ground where fifty years ago not one was to be seen. Numbers appear to be increasing everywhere, and the animals are re-occupying some of their old territories. Commercial exploitation of the fur-seals has not yet begun, but the herds must soon be ripe to yield a harvest. When sealing is taken up again it will be under licence, so that the herds are managed properly to give a crop that will not damage the stocks, and will ensure a plentiful annual supply of seal skins for as long as they are wanted.

Of the four other true seals of the Antarctic, two, the Weddell and the crab-eater seals, live in large herds on floating ice. They have never been seriously exploited, though some unsuccessful whalers in sailing ships made a raid on them in the 1890s. Since then they have been used by man mainly for feeding the working dogs of the numerous exploring parties of several nations that have kept permanent settlements in the Antarctic since the end of the Second World War. Now that disaster has overtaken the great Antarctic whaling industry some commercial exploiters are turning their attention to the possibility of making a profit out of this hitherto untapped natural resource. It is said that the population of crab-eater seals is larger than that of any other kind, and runs into millions, though the statement is based more on informed guesswork than on exact counts.

The main diffiuclty in attacking these seals is their inaccessibility in the vast fields of heavy pack ice; the use of helicopters, however, to take in killing gangs and to bring out the blubber and hides to the sealing ships might solve this problem. Whatever happens, there will not be a mass slaughter to extermination, because any exploitation will be subject to control by the signatories of the

Antarctic Treaty, the most remarkable international agreement ever to have been made. It deals among other things with the conservation of all Antarctic wildlife; and the Powers, though exploring and carrying out research in their allotted sectors, make no rival claims to territorial sovereignty. It has been hoped that the peaceful co-existence now prevailing in the wastes of the Antarctic might lead towards more sensible relations between the nations in less inhospitable parts of the world.

Whales have always been a source of wonder to mankind; their huge size, inaccessibility to observation, and economic value have combined to throw an air of mystery over their nature and ways, and of romance over their pursuit and capture by man. In the old and new stone ages the rare stranding of a whale on the coast must indeed have seemed a gift from the gods to the dwellers near the shore, a gift of unlimited meat and fat to last for many days until even those who liked their meat gamey could no longer stomach it. So highly prized where the products of the whale that all whales, porpoises and dolphins stranded on the coasts of England were, until 1972, legally classed as 'royal fish', and were thus the private property of the reigning monarch.

There are two categories of whales, the 'toothed whales', most of the species having numerous single-pointed teeth, though some have few or none, and the 'whalebone' or 'baleen whales', which have a row of flat horny plates set edgeways along each side of the upper jaw. The toothed whales comprise about seventy-four species ranging in size from the common porpoise four to five feet long to the huge sperm whale reaching a length of sixty feet or more; most of them feed on fishes and squids of various kinds. There are only twelve species of baleen whales; the smallest, the pygmy right whale, is about twenty feet long and the largest, the blue whale, has been recorded up to and perhaps a foot or two over a hundred feet in length. The baleen whales feed mainly on small planktonic creatures – crustacea and molluscs – and certain species sometimes take small shoaling fishes. The planktonic food is filtered out from the sea water by the baleen plates, the inner edges of which are frayed into fibrous strands, which together form a mat or filter-bed to retain the food as the water taken into the whale's mouth passes out between the baleen plates.

All the whales carry a layer of blubber beneath their thin smooth skin; it helps to diminish the loss of body heat to the comparatively cold surrounding water, for whales are warm-blooded, air-breathing mammals. It was the oil that could be extracted from the blubber that made whales valuable to commerce until the end of the nineteenth century – since then other products have become important. In addition, the long baleen plates of the right whale were so valuable that the 'bone' from a single right whale could make an Arctic whaling voyage profitable in the middle of the nineteenth century. The baleen was used for many purposes needing a strong but light springy substance – stays and busks, crinoline hoops and umbrella ribs were but a few of them. Baleen plates were called whale 'fins' in the trade, a term used for centuries, and formerly used to distinguish the baleen whales as 'fin' or 'finner' whales from the toothed whales. The name is now confined to a single species with the alternative book-name of 'common rorqual'.

Fisheries for some kinds of the smaller toothed whales, the porpoises and dolphins, have long been carried on along the coasts in several parts of the world, not only for their oil but also for their meat, which is relished by some people, though to others it seems rather dry and uninteresting unless well sauced. Porpoises were formerly thought to be fish – and still are by the ignorant – and consequently their flesh was much in demand in Roman Catholic countries during the Middle Ages for the tables of the rich, both lay and ecclesiastic, in Lent and on fast-days. Porpoise meat was so highly esteemed that it had a place of honour at state banquets and the feasts of city companies from mediaeval to Elizabethan times. In Europe a coastal porpoise fishery still survives in Denmark, where porpoises are netted during the winter as they migrate through the Little Belt to the warmer water of the North Sea.

Some Pacific islanders, however, have more subtle ways of catching dolphins than netting them. In the Solomon Islands the natives go several miles offshore in their canoes, and when they find a school of dolphins they reach over the side and clash two stones together in the water. The underwater noise drives the dolphins towards the shore and has the peculiar effect that when they reach the beach they stick their heads into the sandy-mud bottom with their tails waving in the air, and without a struggle let the natives pull them on to the beach. In the Gilbert Islands some men are said to be able to call large schools of dolphins to the shore, where the natives kill them for food; but how the calling is done remains obscure. In the Faroes, as formerly also in Orkney and Shetland, the small whale or large dolphin called the blackfish or pilot whale is driven ashore by hunters in boats when large schools come into the bays that run deep into the land.

A more amicable relationship towards man is sometimes shown by dolphins when they fraternise apparently by their own desire. Pliny, who lived in the first century A.D., tells of a boy who used to be taken to school every day across the Bay of Naples on the back of a friendly dolphin. A few years ago a young bottle-nosed dolphin became a tourist attraction at Opononi beach in New Zealand where she associated with human bathers, playing ball with them and even allowing children to ride on her back. Another large dolphin was so well known in New Zealand waters that it was officially protected from molestation by an Order in Council – this was the well-known Pelorus Jack, which for about twenty years up to 1913 escorted ships through French Pass on their way between Wellington and Nelson. It has been said that dolphins have on several occasions helped people in difficulties in the water by pushing them up to the surface – such yarns are not incredible, because dolphins are known to help each other in a similar way, and a mother dolphin prods her new-born young upwards to help it draw its first breath.

The instinctive social behaviour of dolphins seems to be readily adapted to allow friendly relations with man – it is as though the animals treat man as a kind of terrestrial dolphin. However that may be, it is remarkable how easily dolphins can be taken into captivity and trained to perform various tricks for the amusement of spectators and the enrichment of their owners. This amenability probably also stems from a propensity to play, as seen when dolphins race alongside a ship or ride in its bow-wave. Dolphins are now being trained for more sinister purposes by people who are not ashamed to descend to such activities; they are being trained for use in war to do such things as sticking limpet-mines on the hulls of enemy ships, or carrying explosives to act as living torpedoes, and for finding, and attaching retrieval gear to, nuclear weapons lost in the sea.

The larger whales, more remote from coasts on the high seas, seem never to have made any spontaneous friendly advances to man – any advances have been hostile and in the opposite direction. Whaling began with the pursuit of one of the few kinds of larger whales that habitually come close to the land, the northern, Biscay, or Atlantic right whale, also called the Nordcapper. A thousand years ago this right whale was numerous off the coasts of Europe from Gibraltar to the North Cape, and round the Atlantic islands from Madeira and the Azores to Faroe, Iceland and Jan Mayen. It was also numerous on the western Atlantic coasts as yet unknown to Europe. In early mediaeval times the Basque fishermen of the north coast of Spain found that they could capture and kill this rather slow-moving whale with hand harpoons and lances, if they rowed their open boats close to the monsters when they approached the coast. They towed the dead whales to the shore and cut them up on the beach to boil oil out of the blubber, and to remove the baleen and other valuable products. The fishery flourished until half way through the eighteenth century, and centred on such ports as Biarritz, Santonia, Castro Urdiales and others west to Santander. The whalers kept their boats and gear in readiness to put out for the chase when the look-outs in the watch towers on the hills gave them the signal.

The Basques could catch right whales from their open boats because the right whale is one of the few kinds that float when dead. Furthermore the females come into coastal waters to give birth to their calves; they are then very vulnerable because if a calf is killed the mother is reluctant to leave and can be more easily harpooned. This kind of whaling cannot last indefinitely, for exploitation that kills the young before they can grow up and breed is bound to reduce the population. By the beginning of the nineteenth century the Atlantic right whale was scarce and the fishery negligible. Basque whalers did not, however, pursue their trade only in home waters; soon after the discovery of the fishing banks of Newfoundland they were at work catching right whales among the large fleets of fishing vessels from many European countries gathered on the Grand Banks for cod.

In the early part of the seventeenth century the English Muscovy Company's expedition to search for a north-east passage discovered Spitzbergen, which was for long thought to be an eastward extension of Greenland. The explorers did not find a passage but, as mentioned earlier, they did find great numbers of walruses, then called sea-horses, which they slaughtered for their ivory, oil and hides. They also saw great numbers of right whales, not the kind familiar on European coasts, but a new one with a higher oil-yield and longer baleen; this was the Arctic right whale or bowhead, a circum-polar species living in both the Atlantic and the Pacific sides of the Arctic ocean. The Company sent more expeditions to exploit the discovery, and hired Basque whalers to teach them the way to catch whales and boil out the oil. It was unsuccessful in its efforts to maintain a monopoly on the new fishery, which fell into the hands of the Dutch who established a blubber town in the wastes of Spitzbergen on which the industry centred for many years. Smeerenburg was a town of wooden and brick houses, with oil-boiling works, smithies, coopers' and riggers' workshops, storehouses, retail shops, inns and other amenities sought by sailors when ashore. This 'Greenland' fishery flourished for about a century and then, like the old Basque coastal fishery, decayed through over-exploitation.

In the meantime the settlers on the coasts of New England began catching Atlantic right whales from open boats and towing the carcases ashore for dismemberment, a technique they learned not from the Basques but from the native Indians. As the fishery grew they used larger boats that could be away from home for several days, and soon took to decked vessels. These small ships went further offshore to find whales, and came across another kind, the sperm whale, the largest of the toothed whales, which they were able to catch and bring ashore successfully, because it, too, floats when dead. The sperm whale, however, is essentially an oceanic species which does not usually come close inshore like the right whale, so that the whalers who sought it went further and further afield. When one of the whalers invented the hunting method of lowering small boats from a ship to catch and kill sperm whales, stripping the blubber off the carcase lying in the water alongside the ship, and of boiling out the oil in 'try-works' on deck, the great American high-seas whaling industry was born. Now that the whalers no longer had to tow dead whales ashore for treatment, the oceans of the world lay open to their enterprise. In the nineteenth century the industry grew fast, with voyages at first in the Atlantic, but soon extending round the Horn into the Pacific and round the Cape to the Indian Ocean, the East Indies, Australia, New Zealand and the subantarctic islands. Voyages sometimes lasted as long as five years, with the oil produced being sent home by freighters from various ports of call. No place where whales could be found, however remote, was left undisturbed.

Although the sperm whale was the first quarry, other kinds of whales were also hunted. The whalers found that the coasts of the southern hemisphere were the habitat of a right whale nearly indistinguishable from the north Atlantic right whale – the southern right whale, which they harried unmercifully almost to extinction. On the high seas they took pilot whales, which they called blackfish, bottle-nosed whales and other toothed whales such as dolphins of different kinds; in addition they often added sea-elephant and penguin oil to their catches. In the north Pacific they went through the Bering Strait to hunt the Arctic right whale when San Francisco became the headquarters of Pacific whaling.

The decline and fall of the American sailing-ship whaling industry was complex and not entirely due to depletion of the stocks by over-fishing. In the first place the American civil war dealt the industry a grievous blow, for many of the 700 whale ships, nearly all belonging to northern ports, were captured or burnt by Confederate cruisers and privateers, so that owners laid up their ships rather than send them to face the war risks at sea. In addition the northern government bought forty whalers, the famous 'stone fleet', and sank them filled with stones at the harbour mouths of Charleston and Savannah to foil blockade runners and prevent privateers from using the ports. When the industry began to recover after the war it had to meet increasing competition from the quickly growing petroleum industry, and it continued to decline through the rest of the century. In the Pacific, disaster overtook the industry in 1871 when thirty-four of the forty-one fine whale-ships from San Francisco were beset in the ice and abandoned north of the Bering Strait.

While the American whaling industry was evolving, whaling from European ports continued, but on a much smaller scale, chiefly by annual voyages to the Arctic to hunt the bowhead along the ice-edge of the north Atlantic and in the remotest corners of Baffin bay west of Greenland. The Arctic whalers did not boil out the oil on board but brought the minced blubber home in casks to be boiled on shore at the whaling ports. Yet owing to the value of the very long baleen plates of the bowhead, the fishery lingered on into the twentieth century, by which time bowheads were very scarce, largely owing to the practice of killing young calves in order to kill their mothers more easily. The Arctic fishery, like that for the southern right whale, destroyed itself by irrational exploitation and over-fishing, which brought the population of bowheads to such a low level that it no longer paid to hunt them. The end of old-fashioned whaling with sailing ships came with the outbreak of the 1914-18 war; a few ships went whaling for a year or two afterwards, but in the changed economic climate they could not make it pay. It was economic competition and inflated expenses that finished the hunting of sperm whales, not a lack of quarry, for, unlike the right whales, sperm whales were still abundant.

Although the right whales had been hunted almost to extinction the seas still held great numbers of large whales that could not be captured by the old hand methods. These were the large rorquals, the blue, fin, sei and minke whales, and the humpback whale; they are too large and active to be attacked from open boats and, more important, they sink when dead and so would drag down any boat that harpooned them. In the 1860s a Norwegian whaler, Svend Foyn, invented a way of catching them, and thereby started the modern whaling industry. He used a small fast steamship as a catcher, which killed and secured the whale by firing a massive harpoon with an explosive head into it from a cannon in the bows. A stout rope attached to the harpoon ran over a pulley in an 'accumulator', a device with strong springs that allowed some give when heavy strain came on the rope so that it did not break while hauling the dead whale to the surface. When he got the carcase to the surface he pumped compressed air into it to make it float. He made no attempt to boil out the oil at sea, but towed the carcase to a shore factory where steam winches dragged it out of the water to be cut up and processed.

The new technique was very successful, but as it was shore-based, the factories had to be built on coasts where whales could be regularly found not more than about a hundred miles offshore. By the end of the first decade of the present century whaling stations had been built in many parts of the world, from the far north to the Antarctic, in the tropics as well as the colder places; for the whales are migratory, and though they feed on the abundant plankton in high latitutdes they go to warmer waters to give birth to their young.

Some shore stations have worked profitably to the present day, but many were closed after a short life, for various economic reasons. In the Antarctic, however, they were too successful. Up to the end of the nineteenth century little was known of the huge stocks of whales in the Southern Ocean until the Norwegian whalers began exploiting them. By the outbreak of the 1914-18 war many stations had been built, in which inefficient methods wasted much of the catch in order to concentrate on oil production for quick profits. By the end of the war, whales, though still abundant, were noticeably fewer, so that regulations were brought in protecting some species and requiring the use of less wasteful methods.

The ingenious whalers soon found that they could avoid irksome restrictions and increase their field of action, by inventing pelagic whaling. For many years, in places where the building of a shore station was difficult or impracticable they had used floating factories, large ships fitted with machinery like that of shore stations, which anchored in a sheltered bay to process the whales brought in by the attendant catchers during the season. Their methods differed from those of a shore station only in that they cut up the whales in the water alongside, instead of hauling them out on to a flensing plan. In 1925 the 'Lancing', fitted with a slipway for hauling whales on to a flensing deck, was the first to lead the way to pelagic whaling in the Antarctic.

Throughout the 1930s pelagic whalers of several nations rushed to the harvest, but in spite of mutually agreed quotas on the catch, and against the advice of biologists specially employed to study the state of the whale stocks, they annually took more than the sustainable yield. They met with a setback before they had reached the inevitable result of over-fishing – they produced more oil than the market could take, so that they had to agree to refrain from fishing for a season to allow the oil accumulated in previous years to be sold off. After 1945 pelagic whaling re-started with greater intensity, with the result that by the end of a decade stocks were drastically reduced and the largest species, the blue whale, seemed in danger of being exterminated. As the fishing became increasingly unprofitable the whaling companies began withdrawing from the industry, selling their fleets to Russia or Japan. Russia's politically directed economy does not require industry to produce cash profits, and Japan, which has for centuries had a shore-based whaling industry at home, wanted whale-meat to feed her growing population.

Whaling has damaged the whale stocks of the world, and especially those of the Antarctic, so grievously that some countries have recently forbidden the import of whale-products in the hope that the threatened extermination of whales may be avoided. England is among these countries, but the English prohibition is little more than an empty gesture because, although raw whale products are banned, manufactures that include whale products are not, such as dried soups the basic ingredient of which is whale-meat extract, and canned pet foods. A special whale fishery to supply meat for pet foods has grown up in Scandinavia, the 'small whale fishery', exploiting minke whales, killer whales, blackfish and other comparatively small species. The last European-based pelagic whaling with large factory ships after the collapse of the Antarctic industry, was the exploitation of sperm whales off the Pacific coast of South America – but that, like the fishery for the anchovetta, has now come to an end.

The few species of sea-cow, the manatees and dugong, form the order Sirenia, and are in no way closely related to the seals or whales. They are completely aquatic, inhabiting the coastal waters and estuaries of the tropics and the rivers of South and Central America. They are exclusively herbivorous, feeding on aquatic plants of fresh waters and the vascular plants of the Zosteraceae and allied families that grow in coastal shallows, but not, as has been often stated, on algal seaweeds. They are gentle, slow-moving animals that reach a length of eight to ten feet; the body is fish-shaped, the fore-limbs are small flippers, the hind-limbs are represented by internal rudiments not visible externally, and the tail forms a broad, horizontally flattened fluke. The manatees of both sides of the Atlantic and the South American rivers, and the dugong of the Indian and western Pacific oceans, are easily captured with harpoon or spear. They are prized as food by the natives of the countries adjacent to their habitats, such as Ceylon, where in the markets of some coastal towns customers are served with lumps of dugong meat carved from the animal while still alive – living meat does not suffer the quick putrefaction that spoils dead meat in tropical temperatures.

Sea-cows have decreased in number during the last hundred years and are now extinct in many of their former habitats – over-exploitation of the population is hastened by the use of fire arms and the outboard motor engine, which can drive a dug-out canoe as easily as the most highly varnished dinghy. In some countries sea-cows are given at least nominal protection, though the dugong has recently been found to be much more abundant off the coasts of northern Australia than was formerly believed, and the freshwater manatees of some American rivers are still plentiful. Manatees can be seen in the river in the centre of Miami city, where they appear to enjoy swimming in the warm water effluent from a power station; and in Guyana they have served the useful purpose of eating up the water hyacinth that threatened to choke the drainage ditches and other channels. The proposal to introduce them as aquatic mowing machines to other parts of the tropics has not yet been put into practice. Sea-cows have not been heavily exploited for commercial purposes by western man, though the huge Steller's sea-cow of the north Pacific was exterminated by Russian sailors soon after it was discovered in the eighteenth century. The decline in the population of sea-cows, where it has occurred, appears to be due to continued persecution on a comparatively small scale by peasant fishermen.

Many kinds of aquatic mammals, particularly the marine ones, have suffered grievous destruction of their populations at the hands of man, but at the same time give examples of how exploitation should, and should not, be conducted. It is entirely right that man should harvest marine mammals for their oil, skins, meat and other useful products, but obviously foolish for him to expend the capital as well as taking the yield. Harvesting the polygamous seals, such as the fur and elephant seals, is so easily managed to provide a crop without damaging the stocks that even some of the early sealers of the 'gold-rush' days protested at the disgraceful waste of resources which they practised. The seals that herd in immense numbers to breed on floating ice, such as the harp seal of the Arctic and the Weddell and crab-eater seals of the Antarctic, are, like the whales, not amenable to farming methods, so that rational exploitation can best be achieved by international agreement on the numbers to be killed.

After the early years of the present century there were signs that the new whaling industry had already started along the road so disastrously followed by the Arctic whalers and sealers in the last. For the past fifty years governments, especially those of England and Norway, have at great expense conducted scientific research on marine mammals, with the particular purpose of gathering the information about the biology of the animals that could lead to the framing of international agreements for the rational regulation of the sealing and whaling industries.

The investigations, and the rapid decline in the number of whales, led to the formation of the International Whaling Commission to offer guidance to the whaling industry. The Commission had a difficult task in trying to persuade the vested interests to follow advice based on scientific studies of the biology of whales; it had to overcome international rivalries and jealousies, as well as the reluctance of big business to moderate its production for the sake of future prosperity. A system of quotas for the participants in the annual catch was devised, but the sizes of the quotas were always edged up at the negotiating table so that when agreements were signed the allowed catch was far too large. Thus in spite of agreed close seasons, and areas closed to whaling, the decline of the industry went on. Acceptance of the principle of maximum sustainable yield came at last, but too late to save the Antarctic pelagic whaling industry from ruin owing to the scarcity of whales which it had itself brought about. The history of sealing, and of the destruction of the Arctic whale fishery was repeated, although the way things were going had been obvious to all concerned for fifty years.

If the Antarctic whales are left unmolested for a long period of years we may expect them to regain their numbers. Should it then be found necessary or desirable to harvest them, mankind will indeed be crazy not to heed the lessons of the past. The recovery of the whale populations will, however, take a long time. The estimated stock in the Antarctic of the blue whale, the largest animal that has ever lived on earth, was over 150,000 before exploitation began, but by the late 1930s it was about 40,000, and in the early 1950s less than 10,000. In 1962 it was something between 930 and 2,700 – extinction was very near. The size of the stock that would probably best make use of the food available to it is about 125,000, which would give a sustainable yield of 6,000 whales a year. If blue whales were given absolute protection it would take fifty years or more to build the stock up to this size. The other species of whale are in somewhat better plight, but they are travelling the same trail.

5 HAWKS AND DOVES

Birds, unlike the earth-bound mammals, have the advantage of living in three dimensions, an advantage that has allowed them to use a great variety of ecological habitats throughout the world, and to evolve no less than 8,600 living species. Among the mammals one-fifth of the species are bats, which enjoy the same advantage and of which there are some 900 kinds; though they are more limited in their distribution by the nature of their food, which is not universally available. When we go upstairs or climb a tree we cannot claim to use the third dimension in the same way – we merely taste it but do not have it. In the sea we can, with the whales, enjoy the third dimension, which is perhaps one of the reasons for the popularity of Scuba diving – our attempts at flight are not comparable, for even a glider pilot is merely a passenger in his machine; perhaps a Scafa, a self-contained aerial flying apparatus, will one day be invented.

If the number of species is a criterion, the birds have had more evolutionary success than the mammals – but then they have been longer at it, for in the Jurassic some 160 million years ago, when the mammals were small, inconspicuous animals creeping about to catch insects, birds as large as crows were already gliding through the air. The earliest known bird, *Archaeopteryx*, had some reptilian characters such as teeth in the jaws and vertebrae in the tail, though some of its skeleton was wholly avian – and it was covered with feathers. It is certain that feathers came before flight, and served to keep the body warm, just as fur and hair keep mammals warm. It was the structure of feathers with webs on each side of a central quill that made the evolution of the avian wing possible. The plumage, too, hides the shape of the body by giving it streamlined contours; a glance at a plucked chicken shows that birds are merely tiresome little warm-blooded reptiles, disguised in feathers, jumping up and down and chirping. On the other hand the plumage gives birds a beauty of form and colour highly pleasing to human eyes, but whereas a naked bird seems hideously indecent a naked mammal such as a whale, an elephant, a rhinoceros or a man does not. Similarly the song of singing birds is gratifying to human ears, though for the birds it is strictly utilitarian in advertising that a territory is occupied, in sexual stimulation, or in preserving the bond between a pair.

The power of flight gives birds such agility in escaping from danger that most of them are active by day and do not have to preserve their safety by skulking under the cover of darkness. Their universal presence and conspicuousness have ever attracted man's attention, for they and their eggs are edible – some more edible than others – and their plumage and other products are, or have been, valued for various purposes. On the whole, until modern times, man's predation on birds has had little adverse effect, though his modification of the ecological

1. Breton fishermen hooking tunny in the Bay of Biscay.
L. Le Meroy

2. Part of a rookery of fur-seals on an island in the Okhotsk Sea.
Camera Press

3. (above) Natives of the island of St. Kilda catching fulmars about 1920. A long pole with a running noose is gently lowered over the head of a bird sitting on its nest.
Niall Rankin

4. (right) A St. Kildan with his catch of fulmars.
Niall Rankin

5. (top right) Stripping the blubber from a whale on the deck of a pelagic whale ship.
Dr. R. Laws

6. (far right) Cape Gannets nesting on Malgas Island, Saldanha Bay, South Africa. The men are scraping up the guano and bagging it to carry to the landing jetty.
G. Broekhuysen

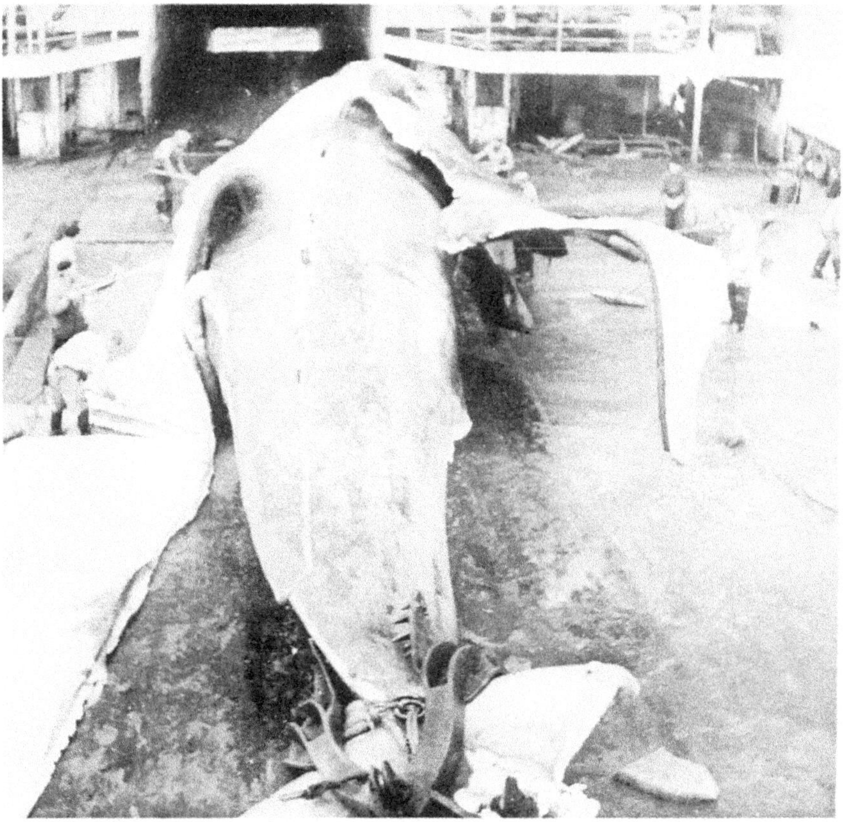

7. Burning unwanted straw in the field after harvest. Much damage is caused to hedges and trees by careless burning.
Aerofilms Ltd.

8. Rhesus monkeys in the Australian Commonwealth Serum Laboratory for making anti-polio vaccine.
Camera Press

9. The large cats have been over-cropped to provide the luxury fur trade. This Dior 1969 Collection includes the skins of jaguar, leopard, cheetah, clouded leopard and ocelot.
Fox Photos

10. Primeval forest in Mato Grosso, Brazil, destroyed by burning to make way for cattle ranching.
The Royal Society/Royal Geographical Society

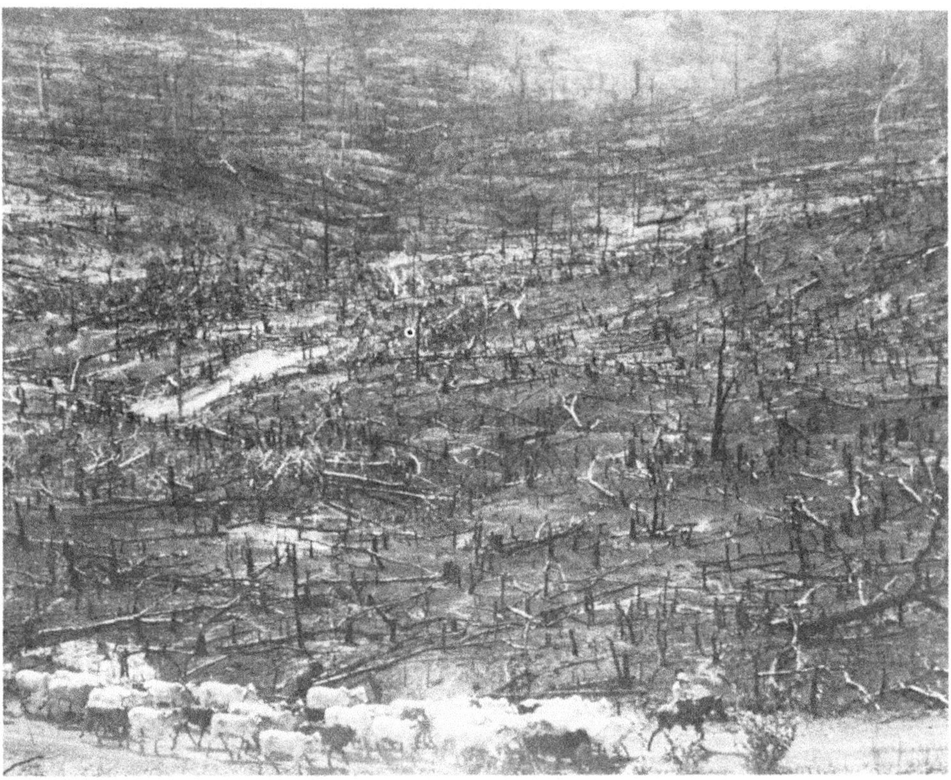

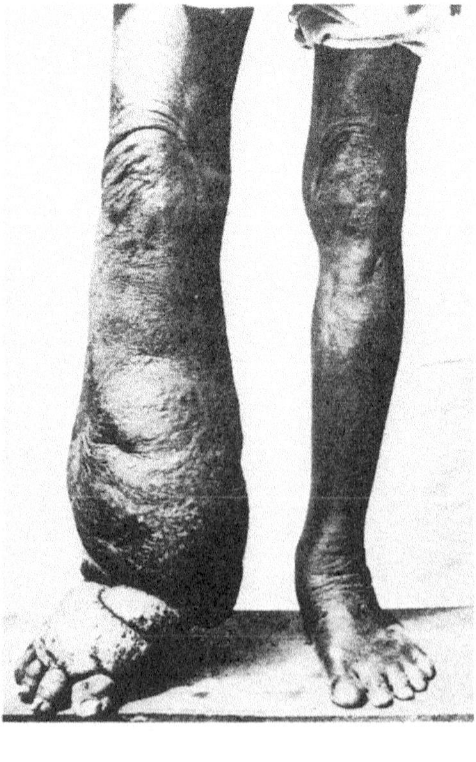

11. A swarm of locusts in Ethiopia.
Gianni Tortoli. Camera Press

12. Elephantiasis of the leg caused by infection with the worm *Filaria.*
The Wellcome Museum of Medical Science

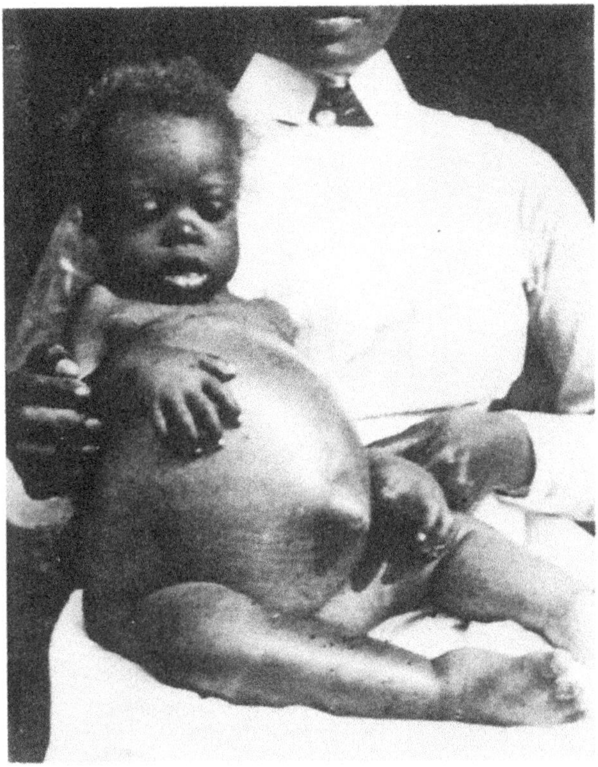

13. An African child infected with hookworms which distend its abdomen.
The Wellcome Museum of Medical Science

14. A potato damaged by one of the several different pests known collectively as 'wireworms'.
Murphy Chemical

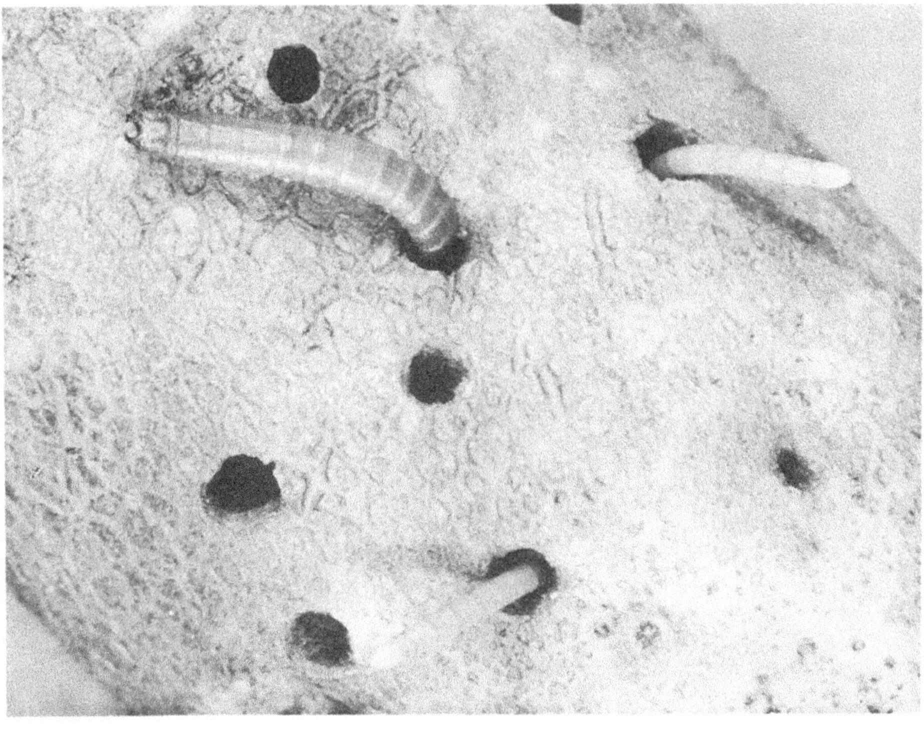

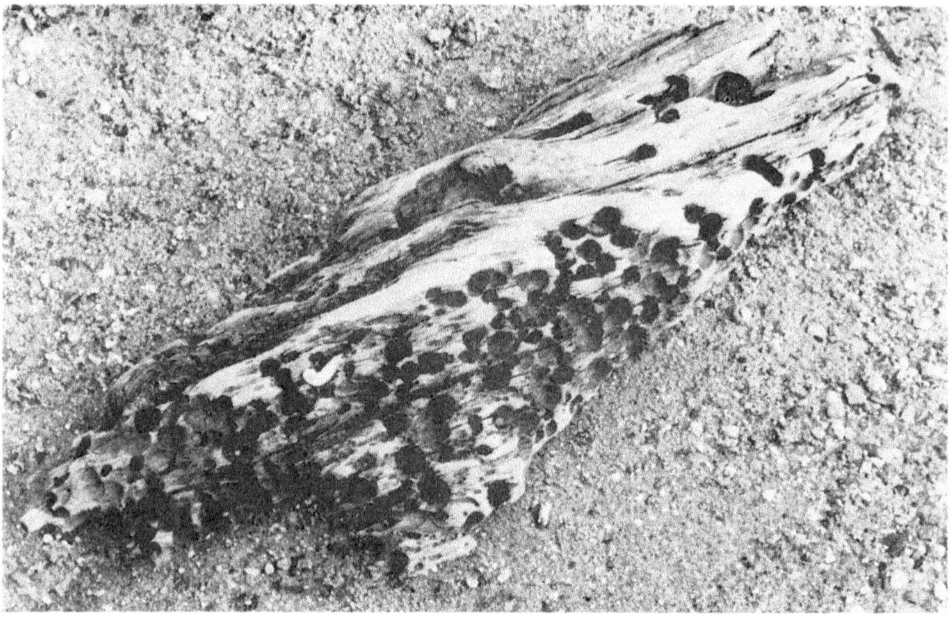

15. Beach-worn driftwood honeycombed with ship-worm (*Teredo*) burrows. One burrow still retains its calcareous lining.
James Cross

16. Man- and sheep-made landscape. The South Downs carpeted with sheep-nibbled turf, and below the scarp the village of Fulking set in a pattern of the enclosed agricultural fields of Sussex.
Aerofilms Ltd.

background by the destruction of forest, and in other ways, must have altered the species-composition of the bird-fauna of any region so treated.

Primitive man's first interest in birds was for food, and he no doubt ate them whenever he could catch them. Bird bones are not numerous in early living-sites, nor in later shell middens. Some bones of the large kinds such as pelicans, geese, swans and cranes, and a few of smaller kinds, have been found, but the bones of smaller birds and the shells of eggs are too fragile to be preserved in quantity with more robust objects. Primitive men were too few and birds were too numerous for man's predation to take more than a sustainable yield. In recent times small groups of primitive people preyed upon wild bird stocks without damaging them permanently. The Eskimos made large catches at bird cliffs and other places where birds gathered in great numbers to breed during the summer, and stored them in deep-freeze for winter use. They made one of their greatest delicacies by skinning a ringed seal through a small cut and filling the resulting skin-bag lined by its own blubber with little auks; they then buried the package in snow until it settled into a sort of oil-soaked brawn.

Until the early years of this century the inhabitants of St. Kilda, the most westerly and remote of the Hebrides, took an annual harvest of fulmar petrels, puffins and gannets without damaging the stocks or driving the birds away from their habitats; similar harvests are still gathered in Faroe and Iceland. The young gannets on Sula Sgeir, the most northerly rock of the Outer Hebrides, have been gathered annually for at least four centuries, and probably many more, up to the present day. Young gannets pickled in salt used to be highly esteemed in Scotland; Dr Johnson tried some during his tour to the Hebrides in 1773, and on asking what they might be was told that they were salted gannets eaten by the Scotch as an appetizer before meals. The Doctor replied bluntly that he had eaten six but he did not feel any hungrier than before.

At the opposite end of the earth the Maoris of New Zealand took an annual harvest of mutton birds – the young of the sooty shearwater, one of the petrels – from their breeding burrows on some of the offlying islands, apparently without reducing the population. With the coming of white men mutton-birding became a commercial venture, and the birds were gathered not for personal consumption but for sale. This larger exploitation is now regulated by law to save the birds from extermination. Mutton birds give stomach oil, which is obtained by holding them upside down over a vessel so that the oil runs out of the beak, another oil which is obtained from the body-fat, and salted meat, which like that of the young gannet is regarded as a delicacy by some people. On the islands off the southern coast of Australia a similar mutton-bird industry exists, though the species harvested is a different petrel, the short-tailed shearwater. Proposals to make a market for canned mutton-bird as a gourmet's delicacy do not appear to have borne fruit, at least in Europe.

Wild birds have not only been used as food but have been killed by man at all stages of his progress from the stone age to western civilisation, in order to use the plumage for personal adornment. Students of human ethology may be able to tell us why there is a universal urge to wear feathers in the hair or head-dress, which reaches its climax in the elaborate millinery made from the feathers of birds of paradise by the natives of New Guinea, the feather head-dresses of the North and South American Indians, or the ostrich-plume full dress of Masai warriors in East Africa. Small feathers from brightly coloured birds are, or were, used for making elaborate feather cloaks in South America, Hawaii and elsewhere, garments made with great skill and infinite patience for the ritual use of chiefs and princes.

The hunting of birds for personal use by primitive peoples before the introduction of fire-arms, and the harvesting of birds' eggs at communal breeding places had, with few exceptions, little adverse effect on the bird populations. The internal trade in bird-of-paradise plumes in New Guinea, although extensive, was not destructive before the coming of Europeans. In New Zealand, on the other hand, the Maoris found more than one species of the giant flightless moas still living when they arrived a thousand years ago, but soon after the first waves of Maori settlement a moa-hunting culture developed so that by the time the islands were discovered by Europeans none of the large kinds remained, though one of the smaller species may have survived into the nineteenth century, and even have been hunted for the pot by European sealers in the South Island. Hawaiians made their feather cloaks from the yellow feathers of the mamo, one of the honeycreepers, but as thousands of feathers were needed to make a cloak and each bird yielded only a few, the species became so scarace that the feather tax had to be paid in the feathers of other species. The mamo was reduced in number to the point of no return and became extinct in 1898; several other species of Hawaiian honey-creeper were exterminated at about the same time.

It was not until the commercial exploitation of birds began – harvesting them as cash crops rather than subsistence crops -- that man's activities became seriously destructive to many kinds of birds. During the seventeenth, eighteenth and early nineteenth centuries, when European adventurers were exploring the remote parts of the world in the search for loot and riches of all kinds, they found many sorts of bright and beautiful birds that were new to them in the lands and islands which they explored. The bird populations of the continents and large islands suffered negligible damage from the early explorers, but those of some oceanic small islands were devastated.

Those islands have been isolated for such long periods of time that their bird and other endemic inhabitants have evolved into forms often far different from the original stock. They have become highly adapted for life in their undisturbed island homes – their particular ecological environment, to put it briefly in the accepted jargon. A specialised and comparatively small fauna or flora is particularly vulnerable to damage when man blunders in and destroys, not only by direct attack but also by unforeseen and unintended effects on the environment – and the early explorers neither knew nor cared about such matters. It was inevitable that some species should become extinct, for they were good to eat, easy to catch, small in total numbers, and unable to adapt to new conditions; they were far along the road to extermination before the arrival of man tipped the balance against them so that they were lost for ever.

The universally known example is the dodo – *doudo* or *doido* is the Portuguese word for mad, simple or stupid – a huge, clumsily-built flightless pigeon inhabiting Mauritius, where it was defenceless against the newcomers, who exterminated it before the end of the seventeenth century. On the other Mascarene islands similar species, the solitaire of Rodriguez, is less popularly known, and on Réunion a third species, of which little is known to science. Sailors probably killed dodos for the pot, though one wrote that they were tough eating; and they brought at least one alive to Europe. More destructive to the dodo than man were the animals he introduced, particularly the pig, against which the flightless ground-nesting birds were helpless. Many other kinds of bird were also exterminated, partly through direct predation by pigs, cats and rats, but also through loss of habitat by the destruction of vegetation by goats and pigs. In the West Indies many species of bird were unintentionally exterminated by the practice of burning the vegetation to clear the land for settlement – records of some species have been preserved, but many must have been lost without ever coming to the notice of naturalists. The disastrous effect on these bird populations was thus due to man's disruptive activities in altering the environment, partly by direct action and partly by introducing and liberating his domestic animals, and the weeds of civilisation.

In later times the commercial exploitation of birds has led to a reduction in the numbers of many species and the extinction of at least one, but on the other hand it has led to the protection and preservation of others. The great auk, a large flightless razorbill, was abundant in the western Atlantic when the Newfoundland fishing banks were discovered. The birds gathered in large breeding colonies to lay their eggs on various islands, where they were completely at the mercy of the fishermen who used them as an easy source of bait. The fishermen themselves also ate great auks; ships from Brittany victualled short with salt meat because they relied on gathering and salting great auks in numbers enough to provide for the voyage. A rookery of great auks must have looked much like one of penguins – indeed, the name penguin was originally that of the great auk and was later transferred to the penguins of the southern hemisphere. A hundred years ago the remains of the dry-stone walls that surrounded the pens where the auks were herded before slaughter could still be seen on Funk Island. Killing the birds in such great numbers when breeding – the hens laid only one egg apiece – was bound to exterminate them sooner or later; very few were left when their last breeding place, a small skerry off Iceland, disappeared beneath the sea during a volcanic eruption in the 1840s.

At about the same time, when the southern sealing industry was decaying, sealers took to killing penguins to boil out their oil. The large king penguin which was abundant on the subantarctic islands, where it gathered to breed in extensive rookeries, was thus exploited, especially on Macquarie Island south of New Zealand, and on Kerguelen. The species has recovered considerably in numbers since the industry died out at the beginning of the present century, but the population is still probably smaller than it was before the sealers arrived. The sealers also used to make shoes with penguin skins, taking the bright orange breasts of king penguins to make the uppers both ornamental and warm; this use of penguin feathers was, however, merely for personal wear, or for presents to sweethearts and wives, and unlike the plumage industry never became of commercial importance.

The rise of the plumage trade in the nineteenth century was one of the most extraordinary, irrational and unnecessary phenomena in the history of man's dealings with birds. The skins of bright-plumaged birds had long been imported in small numbers to embellish the cabinets of the curious, some of whom were serious naturalists. Feathers, too, also used in the extravagant coiffures of ladies in the late eighteenth century, and ostrich plumes became popular as adornment for ladies' hair, for fans and for funeral horses in the early years of the nineteenth. It was about the middle of the nineteenth century, however, that the trade in bird-skins grew to a large size. Brightly coloured bird-skins of many kinds were imported in immense quantities from all parts of the world to London and the Continental capitals, where they were sold by auction to the dealers. The trade flourished for over half a century until the outbreak of the First World War.

There were two main users of bird-skins, the milliners and the bird-stuffers. The milliners, many of whom also styled themselves 'plumassiers', consumed enormous quantities of skins for trimming women's hats with anything from a single feather to complete stuffed birds with glass eyes, together with bunches of artificial fruit and flowers. The demand was so great that in South Africa some enterprising people set up ostrich farms, at least a sensible way of satisfying a foolish demand because the plumes were harvested without killing the birds, which lived to grow some more. All other plumage, however, was obtained by killing the birds in astonishing numbers; a single page from the catalogue of an auction of newly-landed consignments of bird-skins in 1884 includes such lots as '1000 ruby humming-birds; 500 blue merles; 254 parrots, various; 268 Green, with brown heads; and 1300 various.'

The birds-of-paradise from New Guinea were much prized, as were the plumes of two species of white egret, both widely distributed in the Old World, and of one in the New. The long filamentous plumes, which grow only in the breeding season, were called 'aigrettes' or 'ospreys' by the trade, and were obtained by shooting the birds in their nesting colonies. This ruthless destruction of beautiful and harmless birds in their breeding season was a major point in the propaganda of those who deplored the killing of birds to satisfy the dictates of fashion. They were able to arouse enough scandalised public feeling to secure legislation prohibiting the importation of aigrettes and other plumage into England. After the First World War the fashion for wearing feathers died out, a fashion that now seems incredibly ridiculous to western women, who have nevertheless adopted other habits of savages then tabooed, such as painting the face, lips, and the nails of the fingers and toes.

During Victorian and Edwardian times there was a vogue for decorating drawing-rooms with cases and glass shades full of stuffed birds. Many country gentlemen, too, made collections of stuffed birds under the delusion that they thereby advanced the study of ornithology. In those days every provincial town had at least one bird-stuffer's shop, whose properietor not only 'set up' birds brought to him in the flesh, but supplied stuffed exotic birds for domestic ornament or the enrichment of private collections. The bird-stuffers and natural history dealers similarly bought and sold the shells of wild birds' eggs for the collections of those whose acquisitive urges and innate cupidity blinded them to the futility of their childish preoccupation. Killing wild birds and collecting their eggs are now illegal in Great Britain, so that collectors who cannot afford to buy antiques, objets d'art or pictures join the schoolboys in gathering postage stamps, matchboxes and other quaint objects.

The use of feathers for stuffing pillows, bolsters and featherbeds, however, cannot be condemned as it serves a useful and rational purpose. Although the feathers of sea-birds have been thus used in the past, practically all the feathers supplied to the trade now come from domestic poultry, so that wild bird populations are unaffected. On the other hand the use of eider down in making warm bedding is an example of harmless exploitation of wild birds. All ducks line their nests with down plucked from their own bodies, that of the eider duck being one of the lightest, warmest, most elastic and coherent. Eiders nest on the ground on skerries, small islands and mainland coasts from the Farnes and the Scotch border to Scandinavia, Greenland and Iceland northwards to the Arctic, generally in some irregularity of the ground or under the lee of a stone. In Scandinavia and Iceland the nesting grounds are valuable property, and the ducks are strictly protected. The proprietors take the eggs and down at frequent intervals during the nesting season, causing the ducks to lay more eggs than the normal clutch of four to six and to pluck much more down than would be needed for an unharvested nest. At the end of the season the ducks are left a few eggs to hatch, and the final down-lining is taken after the young have left the nest. It is a common practice to make artificial nesting sites to attract the ducks, even to building streets of little unroofed stone cubicles, which are readily accepted. The eggs taken are used for food, their rather strong flavour being much esteemed in the north.

In the eighteenth century the Swedes were not so careful of their natural resources; Linnaeus said that eiders could be seen decorating the fish market in Stockholm throughout the spring and summer. He found that the natives of the Baltic islands used eider eggs for making pancakes, and removed the whale-oil taste from the birds by parboiling them in hay before roasting them. He prophesied that the time would probably come when the excellent down of the birds would save them from being shot.

The gathering of wild birds' eggs for food is economic only with birds that nest in large colonies on or near the ground, hence the species exploited are mainly sea-birds such as gulls, terns and some of the smaller albatrosses. Most birds will lay a second time if their first clutch is taken so that, provided the egging is done judiciously, the breeding stock is undamaged. In some places where egg-picking has been overdone and numbers have declined the industry is now controlled by laws specifying the dates of the open season and the numbers allowed to be taken. In the Falkland Islands the gathering of jackass penguin eggs, and on Robben Island near Cape Town the gathering of black-footed penguin eggs, is thus regulated.

The mining of guano, the sun-dried droppings of sea-birds, mostly various species of cormorant, is another industry which protects the exploited birds. In places where the rainfall is very low a deposit of guano quickly accumulates on the rocks where the birds perch and nest. In the course of hundreds of centuries the deposits reached a depth of nearly two hundred feet on the Chincha Islands off the coast of Peru, so that during the last century a large fleet of sailing ships was engaged in carrying cargoes of it to Europe. These huge accumulations are now all used up, but the annual deposits grow rapidly and are removed at frequent intervals. Guano is similarly gathered on the bird islands off the west coast of South Africa, where the industry is a government monopoly. Some years ago a band of ingenious speculators tried to break the monopoly by building large platforms over nearby mainland sea-beaches, on which the birds perched and provided a harvest. Wherever the conditions are right for the formation of guano the birds that make it are carefully protected from molestation by law, but on the Pacific coast of South America causes beyond human control have recently come into play. Owing to changes of unknown origin in ocean currents the vast shoals of anchovetta fish which formed the main diet of the guanay cormorants, and were the subject of an important commercial fishery for extracting fish-oil, suddenly disappeared. This was a national economic disaster, for the country's chief export was lost; and as for the guanays, unless they can find a new source of food, it is probable that their numbers will fall drastically, with a consequent dearth of guano.

Whatever the fate of the guanays, their numbers are so enormous that it seems impossible that they could become extinct. Huge numbers, however, are not necessarily a safeguard against extermination, as the well-known example of the passenger pigeon of America shows. The passenger pigeon was strongly gregarious, nesting and roosting in prodigious flocks estimated by those who saw them as numbering millions. In colonial days they were killed in thousands to protect the settlers' crops, and even exorcised by a Roman Catholic bishop in 1687. During the nineteenth century they were killed in great quantities for the market, and because they were taken at their nesting colonies their numbers steadily declined until, almost suddenly, the passenger pigeon was a rare bird – the last one died in Cincinnati zoo in 1914. If the flocks were indeed as enormous as they are said to have been, the amount of gunpowder and shot used in exterminating them must have been stupendous – and the American appetite for pigeon-meat equally surprising.

The extermination of the passenger pigeon was not entirely a commercial operation, for the slaughter of the birds was also enjoyed by some as a sport. The line between killing wild animals for food and killing them for amusement is always difficult to draw. In some countries only certain species of birds are regarded as 'fair game', whereas in others anything feathered, from tom-tits upwards, is an acceptable target. In the former, game is protected by law with licencing and close seasons; shooting it has become a 'status symbol', surrounded by mystique and arbitary conventions, backed by large commercial interests supplying the equipment and leasing the shooting rights over the land. In the latter, besides the larger conventional game birds, all manner of small birds are sought, especially when they are plentiful during the times of migration. Here again commerce supplements sport, because great quantities are caught with nets and other devices to supply the market. The English express horror at this trade, which is extensive in the Latin countries, not only because of the destruction of bird life but also for sentimental alleged reasons against eating song birds. They forget that in the earlier years of this century ortolans, wheatears and larks were expensive delicacies in London's fashionable restaurants, and that blackbird pies, as well as rook pies, were country gourmets' treats. 'Petits oiseaux' or 'becca ficos' roasted *à la brochette* with appropriate trimmings such as fat bacon, artichoke and oysters, are delicious – you crunch them up whole like a cat eating a mouse. The natural mortality of birds during migration is enormous, so the toll taken by the bird snarers and shooters may be less destructive to stocks than is supposed.

Where game is preserved, as in England and other countries, the habitat is artificially loaded with a population of birds far in excess of the natural carrying capacity, and predatory or allegedly predatory mammals and birds are ruthlessly killed even if legally protected, so that the whole ecology of the countryside is altered. In 1973 a party of eleven English guns spent a week shooting red-legged partridges on an estate in Spain. The prospects for the season were good because a vast proportion of the magpies and jays had been killed by poisoned eggs. Five days work produced 2,212 head, the best day 610 partridges, one rabbit, 67 magpies and jays, and one pigeon. At least one species of game-bird, the quail, is now bred commercially for the market, where it arrives without running the gauntlet of the sportsman's gun. Doubtless other kinds could be similarly supplied, but why should breeders merely wring their birds' necks when people are ready to pay fancy prices for the pleasure of shooting them when 'turned down'?

In England before the Industrial Revolution, when the population was only some nine million, great areas of the country were still undrained and unenclosed so that huge numbers of wild birds provided a living for professional wildfowlers; but by the middle of the nineteenth century wildfowling was declining. At the same time the rapidly growing and increasingly prosperous human population of the towns was trying to emulate the pastimes of the upper classes on their country estates. Out of the towns emerged the contemptuously named 'cockney sportsman' personified as Mr Briggs in John Leech's caricatures in '*Punch*', and as Jorrocks in Leech's illustrations to Robert Surtees' novels.

One of the amusements of these people when on holiday at the seaside was to hire a boatman to take them under the cliffs where gulls and other sea-birds were nesting, and to shoot them down indiscriminately, often not bothering to pick up the floating carcases from the water. The practice became such a scandal that even Victorian laissez-faire was affronted, so that the first legal protection of wild birds came with the passing of the Sea-birds Protection Act of 1867. Since then further legislation to protect birds has been enacted from time to time, so that in Great Britain all birds and their eggs, with a few exceptions, are now completely protected all the year round. Nevertheless, although a schoolboy has been prosecuted and fined for shooting a blackbird with an air-gun, prowling poachers with illegal weapons still infest some parts of the country in and out of season.

Protection of wild birds by Act of Parliament has undoubtedly led to an increase in the numbers of some species; the outlawing of the practice of taking plovers' eggs for the gourmet trade, for example, has led to a great increase in the population of lapwings, to the embarrassment of aircraft landing and taking off at aerodromes where the birds find the wide undisturbed areas of turf much to their liking. On the other hand some of the modern methods of agriculture are leading to a widespread impoverishment of the bird fauna. The use of insecticides has destroyed much of the food of some species; and the use of selective herbicides has denied many seed-eaters their living. Herbicides, too, have altered the ecological character of much cultivated and meadow land, as have the grubbing of hedges and the burning of straw in the fields – many habitats of small birds are thus destroyed.

Of all the products of wild birds used by man probably the most bizarre are the edible bird nests of the far East. They are made by several kinds of cave-nesting swiftlets, which inhabit south-east Asia, from a sticky secretion of the salivary glands under the tongue. The birds glue the nests to the sides of caves, often in complete darkness far underground, and the bird-nesters collect them at the risk of their lives from flimsy bamboo scaffolding up to a hundred feet above the floor. Some species of swiftlet use straws to make the nest, sticking the materials together with the saliva, but some use nothing but saliva, which hardens as it dries and provides the most expensive grade of nest, highly prized in eastern countries for its supposed aphrodisiac properties. Lesser uses of bird products, such as the former manufacture of quill pens or in the making of harpsichords, hardly affected wild birds because the quills came mostly from domestic geese or feral swans – and the use of albatross wing-bones as long pipe stems or albatross feet as tobacco pouches were merely the diversions of old-time sailors.

In the preceding pages we have examined some of the dealings of man with wild birds, and seen that they sometimes, but by no means always, are to the disadvantage of the birds. There is, however, another side to the picture, the dealings of wild birds with man, often much to his disadvantage; birds can be destructive pests no less than other vermin such as rats, mice, mink or coypu. Some are commensals, living in close association with man and making use of the artificial environment he provides with his buildings and gardens; others avoid his close presence but batten upon his agriculture.

The familiar house sparrow, the 'avian rat', is the commonest of the commensals; it is so closely attached to man that most of its population nests in crevices of his buildings, and few in trees or bushes. It is common in cities, though not so abundant as in the days of horse-transport when it fed upon the seeds in the horse dung that abundantly strewed the streets of even the proudest capitals. The sparrow is a pest to agriculture because it eats great quantities of corn, and to horticulture because it destroys seedlings and tears flowers to pieces. It is bold yet wary, and so well adapted to the artificial environment provided by man that it is abundant in the cities of north America, Australia and in other parts of the world to which it has foolishly been introduced.

The common town-pigeon is derived from the rock dove, an inhabitant of the northern sea-cliffs, which was formerly semi-domesticated and was kept in large dovecots sometimes containing hundreds of pigeon-holes. The birds fed mainly on the crops of the neighbourhood to the advantage of the privileged owner, generally the local squire or parson, who sold the young birds when just fledged – squab pie was once a popular dish. Town pigeons are now feral and where they are numerous constitute a nuisance by fouling the buildings on which they nest, and the pavements below. In some places, as in Trafalgar Square in London or St. Mark's Square in Venice, their large flocks are a tourist attraction, but elsewhere the authorities try to control them. Such, however, is popular sentiment that control with poisoned or stupefying baits has to be carried out in the early hours of the day when few people are about, and on enclosed premises.

The huge flocks of starlings that roost on the ledges of buildings in some towns, especially London, flying in from their feeding grounds as much as twenty miles away, are also unwelcome because they foul the buildings and streets. On a much smaller scale the pied wagtail has a similar habit of roosting in towns; assemblages of several hundreds can be seen during winter evenings on the ledges of some buildings in Dublin. An unexpected new commensal is the kittiwake gull, which has taken to nesting on the ledges of warehouses at the docks of some northern ports. In the tropics the Indian mynah is a commensal with man; it has been introduced into many countries away from its native land, and has become a destructive pest. On the other hand, the swallow, the martin and the swift, which nest in or on man's buildings, are generally regarded as welcome guests, in spite of the mess that martins make when feeding their young.

In addition to the commensals which have adopted and benefited from the artificial architectural environment provided by man, there are many species that, by exploiting another part of his artificial environment, have become pests. Agriculture, growing pure stands of edible plants, provides many birds with an easy living without the need to work for it. In Great Britain and some parts of continental Europe the wood pigeon is the most destructive of them. In Great Britain a large resident breeding population is swollen by huge flocks of immigrants from the Continent during the winter. Wood pigeons are destructive to cereal and bean crops during the spring and summer, and to green crops, especially the brassicas, during the winter. Whole fields of kale or brussels sprouts are often ruined in a few days, and gardens devastated in a few hours. Vast quantities of the birds are shot for the market where they command a good price; and there is said to be a brisk export trade in plucked frozen wood pigeons to France, where they appear as 'partridges' on the menus of the Parisian restaurants. Shooting, however, barely mitigates the damage, for newcomers at once replace the birds removed. An official campaign using stupefying bait has had some success: chick peas, also known as calavances, soaked in a solution of alpha-chloralose are put out for the birds, which become intoxicated shortly after eating them. They can then be taken up and killed, but any other birds inadvertently poisoned can be liberated undamaged after they have recovered.

The rook is another species that is a pest to agriculture, for in cereal-growing districts corn is its staple food – no sooner is the grain drilled in than the rooks descend upon the fields to feast upon it, so that farmers sometimes have to re-sow. They are particularly avid for maize, and in some parts of East Anglia farmers have been unable to grow maize either for fodder or to supply the market for 'sweet corn', because the rooks have eaten every seed sown. The traditional way of controlling them is the farmers' annual rook-shoot at which the young birds, fledged but not yet flown, are shot from their perches on the twigs in the rookeries. They make excellent rook pies. Boys with wooden clappers, rattles, or pop-guns loaded with a few grains of powder, have now been superseded by mechanical cracker-ropes or carbide bangers, which can be set to make minor explosions at regular intervals to scare the birds from newly sown corn, and drive the inhabitants of nearby villages to distraction. The old-fashioned scarecrow is still occasionally to be seen, though it is now generally replaced with empty plastic sacks dangling from stakes and blowing in the breeze. Destructive as the rook is to corn, it is not wholly evil because it also eats about an equal amount of injurious insects – perhaps it may be tolerated in moderate numbers now that the use of DDT and some other insecticides has been made illegal.

In fruit-growing districts the bullfinch and the blackbird are the main enemies. It seems a pity that such a pretty bird as the bullfinch should be so destructive that it is officially outlawed from the bird-protection legislation. In winter and spring it feeds on seeds and buds, and in seasons when there is a poor crop of seeds, or 'keys', on the ash trees it is particularly destructive to the buds of apple, pear and stone-fruit trees. It multiplies its crime because it picks off but drops far more buds than it eats, so that a small flock of birds can strip a fruit tree of its flower buds in a very short time. The best way to combat bullfinches is to trap them in a large cage-trap baited with a living decoy-bird. The blackbird is the bane of the soft-fruit grower; in many districts it is a waste of time to attempt growing soft fruit unless the plants and bushes are protected from blackbirds by a fruit cage.

The damage done to the cockle-gathering industry on some parts of the coast by oyster-catchers has already been mentioned in Chapter 3. On the other hand, some destructive birds are tolerated. Over a hundred thousand gannets nest in about a dozen colonies round the coasts of Great Britain; they are large birds which eat fish, many of them commercial species. If each bird eats only one pound of fish a day – and they must average much more than that – the total eaten in a year is nearly 17,000 tons of fish. Yet gannets are not generally regarded as pests, nor are they destroyed in the interests of the fishing industry.

The damage done by all these birds is, however, nothing to that inflicted on the farmers of subsaharan Africa by the dioch or quelea, a small species of weaver-bird that travels long distances searching for food, in flocks numbering over a million, like swarms of locusts. Queleas are such a serious pest that farmers give up trying to grow grain when they see their crops completely eaten up by the migrating hordes. Governments try to control the birds, but although hundreds of millions may be killed annually by spraying poison onto the roosts from aeroplanes, or burning the breeding colonies with flame throwers, the menace continues. The birds breed at short intervals, and the survivors from control areas fly away to breed and recover their numbers in remote or uninhabited lands. Queleas have always been a pest to native agriculture, but on a minute scale compared with their present depredations – their population explosion has been helped by the opening-up of former wilderness for agriculture, and the construction of new water-holes for the benefit of the exploding human population.

In contrast to these pests some birds have greatly benefited from man's environmental alterations without becoming a nuisance to him. The enormous increase in the numbers of the fulmar, which has been steadily advancing as a nesting species from the Arctic to temperate seas for the last two hundred years, is probably due to the rise of the deep sea fishing industry. All the hundreds of thousands of tons of whitefish landed annually in Europe, America and northern Asia are gutted as soon as they are brought on deck. Their entrails are thrown into the sea, where they cause no pollution for they are at once eaten by fulmars and other sea birds, which thus start the process of re-cycling the waste into the food chain.

Some of the gulls, particularly the herring gull, similarly help in the disposal of waste matter and have consequently multiplied; they not only frequent fishing ports and kipper smokeries to eat the offal, but help to clean the beaches at seaside pleasure resorts of edible litter scattered by holiday makers. In the last fifty years gulls have also taken to coming far inland – the herring, common, and black-headed gulls follow the multi-shared plough undeterred by the noise and exhaust fumes of the tractor, and assemble in huge numbers on town refuse dumps to compete with rats for garbage. Like lapwings, they are less welcome on airfields, where they go to roost and form a menace to aircraft – 'bird-strikes' are a bugbear to pilots. When an airfield is made in the middle of a huge colony of nesting sea birds there is bound to be trouble. On Laysan Island in the Pacific the airstrip was made through the nesting colony of thousands of Laysan albatross, which lay their eggs in shallow nests on the sand. Although hundreds of the birds were killed – 'removed' was the official word – others at once moved into the vacant territories even up to the edges of the concrete runways, and the danger to aircraft remained.

Large populations of scavenging birds are supported unwittingly by man, so that they are far more numerous than they would be without his waste. Crows, kites and vultures clean the villages and even the towns in the tropics, and turkey-vultures recycle the offal of the South American 'frigorificos', the slaughterhouses preparing frozen meat. In tropical Africa the marabou storks, as well as vultures, are officially protected because of the notion that their scavenging prevents the spread of disease – rotting animal matter is certainly offensive to human senses, but bad smells are not necessarily a danger to health.

Birds can, however, spread disease; the most dangerous to man is psittacosis, a virus disease of birds that can fatally infect human beings. The disease was first found in parrots, hence its name, and consequently it was made illegal for a time to import parrots into Great Britain. The ban was lifted later on when it was found that the disease can infect all kinds of birds, and is endemic in some common species including London's pigeons and sparrows. The name of the disease was therefore changed to 'ornithosis'. Human cases of ornithosis contracted from handling wild birds are nevertheless rare, even among those dealing with large numbers such as sea-bird gatherers in the north. Birds may also spread diseases from which they do not suffer themselves. Foot-and-mouth disease of cattle is thought to have been sometimes brought into Great Britain from the Continent by migrating starlings and perhaps other birds, which can carry the virus on their feet but do not themselves become infected. The virus is, however, also carried by many other means, including human boots, and it is now believed to be sometimes carried by the wind.

Psittacosis or ornithosis is dangerous to man mainly because of the popularity of keeping captive cage-birds, a cult that in England is almost confined to imported foreign species. The danger arises from keeping pet birds in domestic living rooms so that infection can easily pass from bird to man. The 'fancy' is now restricted for the greater part to keeping foreign birds, because under the bird-protection laws the ancient trade of the bird-catcher is now illegal, as is the mere possession of caged native species unless they have been bred in captivity. The paraphernalia of the bird-catcher's trade are nevertheless still offered for sale, and it is not illegal to buy trap cages to be baited with a 'call bird', old-fashioned clap-nets, nor the modern mist nets made of fine fine black nylon. The passing of bird-catching has not resulted in a great increase of birds – indeed, the decline in the numbers of many species continues under the influence of environmental destruction, pollution and contamination.

Even people who indulge in the apparently harmless cult of bird-watching, ticking off the names of species in their little lists as they walk about the country, are often unintentional destroyers of bird life, though most of them are ardent conservationists. It has sometimes happened that news of the occurrence of a rare bird has attracted such numbers of enthusiastic admirers that the trampling and disturbance have driven away the very creature they were anxious to welcome. The bird photographers are the equivalents of the collectors of stuffed birds and birds' eggs of the recent past; in their efforts to get good pictures they have often so disturbed the habitat that birds deserted their nests and fled the neighbourhood. The molestation was not always incidental, for removing the vegetation that obstructed the view – an operation known as 'gardening' – could so expose a nest that the birds either deserted it or were exposed to human or other predators. Well might a bird exclaim, 'Save me from my friends!'

Friend or enemy, man's relationship to that part of wildlife represented by the birds is a complicated one. Sometimes it is to his gain and their loss, sometimes to their mutual advantage, at others wholly to his detriment. There is, however, a special relationship apart from all economic and utilitarian considerations, for birds have an aesthetic appeal to more people than other kinds of wildlife. This is quite illogical, for birds are inherently no more admirable than the rest of wildlife, but the fact remains – and the most expert scientific ornithologist is no less a bird-lover than the least of ingenuous amateurs.

6 PRESSURES ON THE LAND MAMMALS

The land mammals' interactions with man are in general much like those of the birds; some have been harrassed to near or total extinction, some have taken advantage of an environment altered by man and have flourished exceedingly, some have become vermin and pests, and others carry diseases which can be transmitted to man or his domestic animals sometimes with fatal results.

Primitive man's first dealings with wild mammals, as with birds, were as a source of food, for a diet of animal protein is both biologically and gastronomically desirable. Nearly all mammals are edible, though some are more palatable than others. In general the herbivores, mainly the rodents and the ungulates, provide the most savoury meat, in spite of prejudice among many people against eating rat or horse, both of which are good food; whereas the flesh- and insect-eaters, the carnivores, bats, and insectivores, provide meat that is less attractive though edible by starving people. There are exceptions: everyone knows that hedgehog roasted in a ball of clay is said to be a gipsy's delicacy, though few have ever tried it, and some people praise the merits of badger or bear hams, which are more gastronomic curiosities than gourmet's pleasures. Carnivore meat is indeed edible, though civilised people have to be reduced to starvation before they will eat dog or cat, and even then they might hesitate to eat lion beef, for a dead lion stinks worse than a dead hyaena.

In the early hunter-gatherer stage of the evolution of human civilisation men were too few and mammals were too many for man's predation to have any adverse effect on the numbers of wild mammals. Some archaeologists have suggested, as we noted in Chapter 1, that in the later stone age man did make an 'overkill' in excess of his immediate needs and that he thus contributed to the extinction of some species – even the mammoth – but the proposition is by no means widely accepted. The domestication of animals was a revolution that deeply affected both man and the wild mammals for, together with the domestication of plants as crops, it is the foundation of our present civilisation; notwithstanding all his achievements in the arts and sciences man must eat and daily fill his belly.

It was the invention of agriculture and animal husbandry, as discussed in Chapter 2, that made human settlements possible first as villages and then as towns, so that urban civilisation could emerge, with division of labour and the growth of specialised trades. The concentration of population in the settlements then came to be fed with the surplus of the surrounding country, made available by the invention of farming. The important effect upon the wild mammals was not so much an easing of the predation pressure from man, as a profound and ever-increasing change in the environment with all its ecological consequences. The wild herbivores became pests, and the carnivores vermin, from whose appetites man's crops, flocks and herds had to be protected by destroying as many of the offenders as possible. This fight led to the extermination of many of the larger carnivores in some parts of the world – the last wolf, for example, was killed in England about 1500 – but has not succeeded in conquering the herbivores, which continue as destructive pests to the present day.

We can only guess about the beginning of the domestication of animals, though it is established that the dog was probably the first species to take up with man, and was followed by the sheep and then by others. The carnivores, the dog and the cat, gave themselves up when they entered into partnership with man, whereas the herbivores, the sheep, goats, cattle, horses and the rest, were enslaved and brought into involuntary subjection. How can wild mammals have been brought into such close relationship with man so that they not only provide him with plentiful food but can be made to slave out their lives as beasts of burden? The answer is probably provided by the phenomenon known as 'imprinting', which is found in the young of both birds and mammals. Young animals become imprinted with the image of their parent at an extremely early age, so that they recognise her as mother; experiments have shown that some kinds of young animals can become wrongly imprinted if they first see another animal or even inanimate objects. It is, for example, practically impossible to tame the puppies of a bitch that has 'gone wild' if they are captured after their eyes open and they have become imprinted with her image; but if they are taken before their eyes open so that the first animal they see is their human captor they can be domesticated.

It is interesting that domestic animals have, in effect, to be domesticated with each generation, so that they are imprinted not only with the image of their own species, but also with that of man, and thus have a double dependence. The young of domestic animals that have gone feral are nearly as intractable as truly wild animals, though in the course of thousands of years they have been artificially selected for docility. There is possibly, however, some genetic basis for aptitude to domestication, probably linked with other characters such as coat colour, for domestic breeds tend to have patterns and types of colour different from those of their wild ancestors, patterns that would be disadvantageous to survival in the wild. Yet in spite of any such genetic bias they must be habituated to handling by man from birth; horses must be broken-in to work because they inherit nothing of the training learnt by preceding generations – and not every lamb goes meekly to the slaughter.

Domestic animals cannot be excluded from a discussion about man and wildlife because man and his animals have drastically changed the environment to the advantage of some animals and to the detriment of others, – they are all part of the complicated ecological web. The destruction brought to the fauna and flora of oceanic and other islands by the introduction of alien species, especially domestic goats, pigs and cats, has already been mentioned, but alterations on a large scale have also been brought about nearer home. The maquis country of the Mediterranean region, dry, rocky and with a sparse vegetation of bitter and aromatic shrubs and herbs, has probably been produced by the action of goats, which have destroyed a more luxuriant plant and tree cover, with consequent soil erosion, desiccation and reduction of rainfall. Most of the countryside in the British Isles is a sheep-landscape; until the enclosures of the late eighteenth and early nineteenth centuries sheep grazed on open range, clipping the vegetation short and preventing the regeneration of woodlands by eating the seedlings – they were probably the main cause of the clearance of woodland from the downs of England and the hills of Scotland. Sheep have been influencing British wildlife, by moulding the landscape in this way, since neolithic times when the small turbary sheep began the process which was later carried on by the improved breeds.

On a world-wide scale few mammals have been exterminated by man in historic times; although some have disappeared from civilised countries they still live in remoter places. Bears, wolves, beavers and wild boars have long been exterminated in Great Britain but still survive, though in reduced numbers, in parts of continental Europe. The two large ungulates described by Julius Caesar in his commentaries on the Gallic war, the urus and the bison, no longer exist as wild animals; the first is extinct and the second lives only in semi-domestication. The urus or aurochs, the large wild cattle with huge bulls sporting enormous handlebar horns, which was probably the ancestor of some breeds of domestic cattle, lived in eastern Europe until the seventeenth century when the last was killed. An engraving, said to be copied from a now lost painting, purports to show its appearance in life. An ingenious German zoo proprietor claims to have 're-constituted' the aurochs by crossing various breeds of cattle, including as an important ingredient the Spanish breed of fighting bulls. He has succeeded in producing an animal that looks much like its alleged portrait and agrees with the descriptions of its conformation and colour, but lacks its great stature, being no larger than the old domestic long-horn – another breed that went into its making. Its genetic make-up cannot be the same as that of the long-extinct model.

The European bison, taller and less shaggy than the American species, lived in the forests of Poland and the Caucasus until the First World War after which only a remnant of the population was left in the Bielowitza forest of Poland. These survivors were reduced to a very small herd during the Second World War, after which they were cherished by conservationists and have increased slightly in number. But none is truly wild; they are fed and tended by their keepers and each is recorded by name in the stud-book – on the principle of not having all the eggs in one basket zoos and animal parks in various parts of the world hold small stocks in the hope of preserving the species.

The wild horse that formerly roamed the open plains of temperate Europe and Asia also now exists only in captivity, except for a small population in Mongolia which is probably no longer a pure strain, owing to contamination with the genes of domestic horses. The wild horse of Asia, the tarpan or Przewalski's horse, is a small stocky animal with coarse head and hog mane, dun in colour with a mealy mouth. Small horses closely resembling it are found in northern countries – and carvings of the stone ages show that it was one of the food resources of our primitive ancestors. Wild horses existed in eastern Europe until about the end of the eighteenth century, though latterly probably much interbred with domestic horses.

An extraordinary myth has grown up about these western horses, holding that they were a species different from the once widely distributed wild horse of Europe and Asia. Enthusiasts supporting this error apply the name 'tarpan' to the supposed European species; 'tarpan', however, is the Kirkghiz Tatar name for the wild horse of Tartary, that is, for Przewalski's horse. The German zoo-man who claimed to have re-made the aurochs also claims to have re-made the so-called tarpan by cross-breeding Przewalski's and various domestic horses. His product resembles a rather large Przewalski's horse with a greyer coat, but as the tarpan never existed in Europe as a different species from the Asiatic tarpan, the reconstruction is not convincing. The Przewalski horse, however, is alleged to have a different number of chromosomes from the domestic horse, 66 as against 64; nevertheless systematists still classify it as a subspecies of the domestic horse *Equus caballus przewalskii*, though it is the ancestor of the domestic race.

Many kinds of mammal are so reduced in numbers that they are in danger of extinction; for the greater part they are either animals of commercial value or targets for the sportsman's rifle, and most of them are of comparatively large size. Some, such as the blue-buck and the quagga of South Africa have already gone. The blue-buck, or blaawbok, was similar to the still existing roan antelope but differed in coat pattern and colour, which was a bluish grey. It was hunted to extinction by about 1800, shortly after it first became known to European naturalists; the rapidity of its disappearance was due to its restricted distribution in the district about Swellendam, so its numbers could not have been great before the settlers came.

The same cannot be said of the quagga, a zebra with striping confined to the head and fore-quarters, which was plentiful on the South African plains from the Capte to the Orange Free State. Settlement and cultivation early drove it from Cape Colony but it remained plentiful further north until after the middle of the nineteenth century. It was exterminated by the Boer farmers who used the meat for feeding their labourers and made money from the sale of the hides which were in great demand at the Cape for making leather – hides were trekked down to the coast by the waggon-load. A survivor lived for 19 years in the London zoo until 1872, and the last known living quagga died in the Berlin zoo in 1875. The London quagga was a mare, but for the six years from 1858 to 1864 a quagga stallion also lived in the zoo. They were never mated, and when the stallion was shot after accidentally breaking his leg the last chance of preserving the species by breeding it in captivity was lost.

The extermination of Steller's sea-cow in the western part of the Bering Sea was mentioned in Chapter 4. Like the blaawbok it was an animal of restricted distribution and consequently of low numbers; it was discovered in 1741, and was exterminated in a little over twenty-five years by Russian traders and fur hunters who found it an easily obtainable source of palatable meat. In the West Indies several kinds of smaller mammals were exterminated soon after their discovery by Europeans, so that little is known about them. They were used as food both by natives and settlers, as is shown by the presence of their bones in middens and occupied caves. Several species of larger shrew-like insectivores lived after the coming of Europeans, but they were exterminated so long ago that no vernacular names for them have been preserved. Similarly two or three large rodents related to the guinea pig were exterminated before they became known to European naturalists, as were some of the island races of another medium sized rodent, the hutia. Thousands of miles to the south, in Patagonia at the tail of South America, remains found in caves suggest that the giant ground sloths may not have become extinct until historic, though pre-columbian, times, and that they may have been kept in captivity as domestic animals by the natives; but there is no evidence that they were exterminated by man.

Many of the unique marsupials of Australia became so scarce after European settlement and the opening-up of the country, that they have long been thought extinct; but from time to time naturalists are surprised by finding some of them still living in out-of-the-way places. Leadbeater's possum, a small bushy-tailed species, was thought to be extinct until in 1961 a colony of the animals was found living in Victoria. Another small possum, *Burramys parva*, was known only from some fragmentary fossil bones but in 1966 was discovered to be a living species. Now that the Australians have woken up to the value and interest of their fauna they protect most of their native mammals, many of which are increasing in numbers.

The fur trade formerly recklessly exploited the species yielding valuable pelts, and sportsmen wantonly killed enormous quantities of animals; protection has reprieved many from extermination, such as the koala, the duck-bill, and others, though it came too late to save some such as the toolach wallaby, which was exterminated by a well-meant attempt to transfer those remaining to a new sanctuary. The killing of large kangaroos for the canned pet-food trade still goes on, but it is to be hoped that it will be regulated to produce a crop that is not damaging to the stock. The early settlers not only killed the animals for fur, food or sport but also deliberately tried to exterminate some species that were regarded as pests. Little success has been achieved in exterminating the dingo, which is not a marsupial and was perhaps introduced by the aborigines some thousands of years ago, but in Tasmania the carnivorous thylacine or marsupial wolf was until recently thought to be extinct. It preyed upon the settlers' sheep and fowls so that a bounty was given for each one killed, and was still being given in the first quarter of this century. It has recently been found that a few thylacines are still living. Similarly the aboriginal human inhabitants of Tasmania were exterminated, at least in part deliberately; the last died about a hundred years ago and only some bones and photographs remain as a record of these primitive people. The deliberate extermination of human populations has taken place elsewhere, as with the Amerindians of Patagonia and Brazil; it continues to the present day under the dignified name of 'genocide'.

The Antarctic wolf, a species of long-legged fox resembling a small coyote, and peculiar to the Falkland islands, became extinct in the 1870s by being killed for its fur and poisoned by settlers to protect their sheep. The 'wolves' showed no fear of man and would approach the camp-fires of the gauchos who, as Darwin saw, attracted them by holding out a piece of meat in one hand while holding a knife ready to stick them in the other. On the island of Chiloe Darwin found another species nearly as trusting, for he was able to walk up quietly behind one and 'knock him on the head with my geological hammer'. He brought its skin home and had it stuffed.

Apart from the species that have been lost for ever there are some that now live only in captivity or as domestic animals. When the few remaining wild Przewalski's horses are gone the horse will join the one-humpted camel, the llama, and the domestic ox in the latter category. In the former there are several species that, though not domesticated, live only in captivity. The peculiar Pere David's deer, which appears to wear its antlers back-to-front, lived only in the imperial hunting park near Pekin until the Boxer rising when most of them were killed. A few were brought alive to England from which a breeding herd was established at Woburn park in Bedfordshire – stock from this herd is now held in many zoos round the world. In South Africa some of the big game antelopes are no longer truly wild animals – the bontebock, blessbok, and the white-tailed gnu live only within fences on certain farms and reservations where the owners crop them to supply the venison market of the cities. A similar fate to some extent awaits all the larger animals of Africa as the growing human population leaves less and less room for wild animals; some of the national parks are already being wholly or partly enclosed with game-proof fences. By the end of the century it is probable that all the larger wildlife will be thus contained for the benefit of the tourist trade.

Finally there are some extinct mammals that never lived – the tarpan of Europe, which was not a species differing from the once common wild horse, has already been mentioned. A similar mythical animal that has been exterminated was the couprey or Cambodian wild ox, now known to be a hybrid between domestic cattle and the wild and well-known banteng.

Although comparatively few species of mammal have been exterminated in historic times a great many have become so scarce that their existence is in danger unless they are deliberately protected. The list of 'endangered' species is long – far too long for mention of more than a few to be made here. It is, unfortunately, made to appear longer than it actually is by enthusiastic conservationists who injudiciously include the names of numerous sub-species in an attempt to make our flesh creep. No doubt this hoodwinks the innocent well-wishers with money to give away but without much knowledge of animals, but it does not inspire respect for the conservationists' methods among those who know, much as they may wish to see wildlife preserved.

Every continent except the Antarctic harbours some endangered species; in Australia, as already mentioned, those which are seldom seen sometimes turn out to be more plentiful than they were thought to be. Similarly in Madagascar most of the numerous species of lemur, including the aye-aye, a species with front teeth resembling those of a rabbit and used for tearing open tree branches to allow its excessively long and thin third finger to fish out insect grubs, have been much reduced in number; some may be extinct, but the standing of most of them is not well known. Their decrease appears to be due more to the destruction of forests than to direct killing by man.

Monkeys in both the old and the new worlds are now subject to heavy predation by man to supply the wants of research laboratories throughout the world, and for the manufacture of certain vaccines by the pharmaceutical industry – one monkey had to die to provide every four doses of the first polio vaccine to be marketed. The huge trade in living monkeys to provide subjects for physiological and behavioural experiments is leading to the capture of less easily obtained species as the commoner ones become scarcer. Of the large anthropoid apes the chimpanzee has been the chief victim for experimental work, whereas the threat of extermination to the orang-outan and the gorilla have been mainly due to the demand by zoos for living specimens. Both the last two species are now protected, but illegal capture in their native countries is difficult to prevent; the prohibition of their import, except under strict export licence, into the receiving countries has been more effective. Before the impact of science on the world's population of monkeys man's predation was small – some were captured for zoos and the pet trade; a few, such as the colobus, which has long silky hair in contrasting black and white, were killed for the fur trade; some were destroyed as agricultural pests; and in some places they were used for human food. The flesh of the lowland gorilla, too, is much esteemed as food in some parts of west Africa where the introduction of cheap firearms has led the natives to destroy far more than they did formerly with spears and clubs; gorilla is said to be delicious, 'nearly as nice as a man' – for there are still surreptitious cannibals who can make a critical appreciation.

The carnivorous mammals have always been persecuted by man, partly because they are predators on domestic animals and sometimes on man himself, and partly because most of them are fur-bearers with beautiful pelts of great aesthetic and commercial value. The trapping of the smaller carnivores such as mink, marten, sable, wolverine, fox, wolf and many others has long been a traditional occupation in northern countries. It still continues, so it is evident that though numbers may be reduced the animals have by no means been exterminated. The modern introduction of fur-farming to breed animals for their pelts has increased the supply and possibly reduced the pressure on wild populations. But it has had other repercussions: great numbers of the smaller kinds of whales are killed to provide food for the animals on fur farms; and lasting effects may be expected from animals that have escaped from the farms, such as the mink in Great Britain, which has introduced a new predator to alter the ecological web of the fauna, and a new pest to the game-fish rearer.

The larger carnivores, the brown and polar bears, the tiger, leopard, jaguar, cheetah and so on have been looked upon as dangerous predators to be exterminated, as the source of valuable skins for luxury clothing and rugs, and as suitable quarry for sportsmen – the lion alone of the big cats has a short-haired pelt that is hardly worth preserving, and is the least endangered species. The beautiful fur of the other cats has been their downfall, so that over-hunting to supply an insatiable and lucrative market has greatly reduced their numbers. Fifty years ago India had more than 40,000 tigers, but since the subcontinent became independent an orgy of wildlife destruction by the masses has reduced the number to less than 2,000 today. Among the bears, both the brown and the polar bears have declined in number through the efforts of the fur-traders and sportsmen. The brown bear, once common throughout the northern parts of the world, is now scarce in most places, for bears, especially the great grizzly variety, are not compatible with human settlement. On the other hand the American black bear, a smaller species, though banished from more settled regions, is still common in extensive woods and forests. In some of the national parks black bears are encouraged to forage in the garbage dumps of hotels for the diversion of visitors. The huge white pelt of the polar bear has always been prized for rugs and trophies, but since the modern opening-up of the Arctic and the use of aeroplanes and helicopters by sportsmen for shooting bears, numbers have declined rapidly – polar bears are now given complete protection in the Russian Arctic.

Most of the ungulate mammals and some others that come into the category of 'big game' have decreased in number as the growing human population steadily encroaches on the wilderness that was formerly left for wildlife. In the early days of settlement in South and East Africa professional hunters made their living by shooting wild animals for the commercial value of their hides, for their meat dried into biltong and, most important of all, for the ivory from elephants. Game preservation laws came into force as settlement spread, so that the hunter's calling died out during the first quarter of the twentieth century. Professional hunters turned to serving as guides for sportsmen, who collect trophies but are unconcerned about the commercial value of their victims; they became the African equivalent of the pseudo-cowboys who wet-nurse the tourists on American dude ranches. Fortunately for the animals, the tendency has in recent years been away from shooting towards photographic 'safaris' – safari, a word which seems to have some magic meaning to tourists, is merely the Swahili name for a journey or caravan, derived from the Arabic.

The East African game laws charged visiting sportsmen high fees for licences to kill game – residents were treated more generously. For some reason not readily apparent the licence to kill a giraffe was so high that few bought one, so the giraffe is still a common animal; furthermore few of even the most opulent sportsmen owned baronial halls big enough to accommodate the stuffed head and neck of a giraffe without looking ridiculous. On the other hand the high fee for an elephant was no deterrent, for many sportsmen even now reckon to recoup much or all of the expense of their safari with the ivory they collect. The licensed hunting of elephants, however, has had no adverse effect on the elephant population in recent years; indeed, the increase in the number of elephants and the spread of settlement have made necessary an extensive annual 'control' of elephants, in which some thousands are killed by government servants.

After the end of British rule when the East African countries achieved independence the illegal killing of game increased enormously. The highly lucrative trade in poached ivory and rhino horn is said to have been condoned or even encouraged by bribery and corruption in high places of the native administration. In 1973 elephant-poaching for tusks had reached an annual total of about 12,000 animals in Kenya, and became such a public scandal that the hunting of elephants even under licence was officially banned. Shortly afterwards Tanzania prohibited all hunting and live-trapping. It remains to be seen whether these gestures towards European sentiment can be enforced in the face of potential high profits among people who think that the products of their own country are their own property.

In the Tsavo National Park a long drought drove the elephants to destroy their environment by tearing up the trees in their search for food and water, so that it was reduced to an arid waste, and thousands of them died. A number of years with normal rains followed, the environment and the elephant population recovered, until another long drought brought another crash. During the first drought 'control' measures to kill the excessive numbers of elephants were proposed, but the control produced by the natural climatic cycle showed that artificial management may be a waste of time and effort. Similarly the large population of hippopotamuses in the Queen Elizabeth National Park led the authorities to take control measures because the animals were destroying the environment. Thousands were killed, greatly to the benefit of the native meat markets, but it has been suggested that a normal cycle of build-up and crash might have solved the problem without human intereference.

The sale of the hippo meat was an incidental side effect of the control operations, but in parts of East Africa, Rhodesia and South Africa some farmers are experimenting with game-ranching on the principle that native game animals are better converters of native vegetation to useful meat than are cattle of European or Asiatic origin, and are less vulnerable to indigenous animal diseases. Their main difficulty is in getting the meat to distant markets; the animals are scattered over wide areas when they are shot, for they cannot, like domestic cattle, be sent to the slaughterhouse on the hoof. Game-ranching could be a means of saving some species from extermination – when the preservation of wild animals becomes commercially profitable, as in ranching and in tourism, the future of the animals is brighter.

The state of the mammalian fauna of Asia is poor as a result of man's activities. The human population explosion, wars, and the relaxation of law and order on the withdrawal of European rule from India on the attainment of independence, have greatly reduced the habitats available to wildlife, and have increased human predation, formerly called poaching. The scarcity of tigers in India has already been mentioned; the once plentiful blackbuck and swamp deer are going the same way, and the Indian rhinoceros has now only a minute population restricted to a few protected reserves. The Sumatran and Javan rhinoceroses of south-east Asia are nearly extinct because of the high prices paid for their horns – which are supposed to have aphrodisiac properties – in the eastern drug trade. The African black rhinoceros is persecuted by poachers for the same market, though the damage done is probably less than that inflicted by the British Government when it had over 900 black rhinoceroses shot and left to rot during the land-clearing in Tanganyika for its abortive groundnuts scheme – a slaughter that was understandably hushed up. On the other hand, the once very scarce white rhinoceros has multiplied so much in protected reserves that there is enough of them to supply the world's zoos – at very high prices. The ungulates of the east, the Malayan tapir, the wild cattle, both gaur and banteng, the wild water buffalo, the various antelopes and deer are all much reduced by the pressure of human settlement and by shooting.

In North America man's attitude to wildlife has greatly altered during the last fifty years. Over great areas much of the larger fauna has been wiped out by settlement and hunting, but now that public opinion has been aroused, protection and the sanctuary of national parks have saved most species from extermination, in spite of the traditonal right of Americans to free hunting – a survival of pioneering days. Hunting, which includes shooting anything from little squirrels upwards, is controlled by licence and close seasons, so that those who want to convince themselves that they are tough guys and wish to fill their 'trophy-rooms' with silent witnesses to their prowess, are unable to exterminate their country's wildlife. Protection came just in time to save the American bison or buffalo; the enormous herds of the prairies were shot for their hides and tongues, and for sport, but during the Indian wars the government deliberately tried to exterminate the remnant so that the Indians, whose livelihood depended upon the buffalo, should starve to death. Similarly the pronghorn, the American 'antelope' which was brought to a very low level, has partly recovered its former abundance. Protection can, however, be overdone; in many places the destruction of wolves and coyotes to protect deer resulted in such an excessive growth of the deer population that the habitat was badly damaged and great numbers of deer died in severe winters.

Wolves have long been banished from settled regions, but are still plentiful in the north where their predation helps to prevent a similar population crisis among the caribou. The coyote is much more widely distributed and in some States is still regarded as vermin to be trapped, shot or poisoned – it seems to have established a relationship to man similar to that of the fox and the badger in Great Britain.

In South America, where such protection laws as exist are not easily enforced, wildlife is the victim of commercial exploitation and sport, so that the pampas deer of Argentina, for example, is now probably lost. In the Andes the vicuña, a wild relative of the llama, and the chinchilla, a rodent rather smaller than a rabbit, have both been reduced to remnant populations by hunters who kill them for their soft, dense fur. Although nominally they are strictly protected, poachers smuggle the pelts into countries from which they can be legally exported to the fur-markets of Europe. The chinchilla, however, is now raised in captivity for the fur trade, and the price of the once highly expensive pelt has fallen with its loss of scarcity-value.

The biggest threat to wildlife in South America, however, is opening up of the northern forest region of Brazil by the building of a motor road through Mato Grosso to the Amazon in order to bring the enormous area of almost unexplored country into agricultural production. As settlement spreads on each side of the road the forest is felled and burnt, so that the whole face of the countryside is changed and the old ecological web is destroyed with its wildlife, from the lowest invertebrates up to the highest mammals – the unfortunate Amerindians, who have recently been deliberately shot, poisoned by sugar laced with arsenic, and bombed from the air to make way for civilisation. Some people fear that the loss of the forest will bring climatic changes that may have harmful results not only locally but on a world-wide scale. They suggest that the loss of photosynthesis will heighten the carbon dioxide content of the whole atmosphere, with a consequent increase of temperature and aridity through the atmospheric 'greenhouse effect'.

Although much wildlife has suffered badly at the hands of man, some has greatly benefited and has found the altered environment much to its liking. Such species among the mammals are mainly the herbivores, in contrast to the universally persecuted carnivores. They are, for the greater part, species that not even the most dedicated conservationist wants to protect, for they are those that have taken advantage of the works of man to become pests. Conspicuous among them are the rodents, which breed very quickly and survive in enormous numbers when inadvertently provided with almost unlimited food by the products of man's agriculture. Rats and mice swarm in the human environment all over the world in spite of the unending war which has been waged against them for hundreds of years; they even defeat the achievements of modern scientific control.

The population of rats, as mentioned in Chapter 2, has been much reduced in recent years by the use of warfarin, which at first appeared to be the ideal poison but is becoming increasingly ineffective owing to the spread of strains of rats immune to its action. New poisons being introduced show great promise, but no-one can tell how long they will remain effective, for the possibility of immune strains of rats being produced cannot be ignored. Some very powerful rodent poisons are known, such as sodium fluoroacetate and fluoracetamide, but they are so poisonous to all other animals and man that their use must be severely restricted; other powerful poisons are so cruelly painful in their action that they are condemned on humanitarian grounds. It seems that the war against pests must go on, but that there is little probability that it will be won, at least by man.

Mammals that damage standing crops are almost as destructive as the pests of stored products. Until myxomatosis brought some control over numbers in Europe and Australia the rabbit caused annual losses running into millions of pounds – but the rabbit was introduced by man to Australia and much of Europe, its natural habitat being the Iberian peninsula. Artifical introduction of animals into new habitats usually results in disaster, for they may find unoccupied niches in the ecosystem or compete with and oust native species, thus upsetting the whole ecological network. In addition to the rabbit, the grey squirrel introduced into Great Britain from America has become a destructive nuisance that has defeated official efforts to exterminate it; the red deer introduced into New Zealand put the country to enormous expense in attempts to reduce their numbers; and the possum introduced there from Australia has also become a pest.

These were deliberate introductions, but unintended introductions can be equally harmful. Rats and mice originally came from central Asia but have been accidentally spread by man throughout the world; other accidental introductions are due to the escape of captive animals from fur farms. The addition of the mink, a carnivore, to the British fauna has already been noticed; similar escapes set the musk rat and the coypu at large. The musk rat was fortunately exterminated soon after it became established in the wild, but only after the expenditure of much time, effort and money. The coypu, on the other hand, is still with us in spite of the destruction of many thousands of the animals in the marshes of Norfolk and Suffolk. Introductions made to control previously introduced pests can be equally damaging though made with the best intentions. As we have already seen in Chapter 2 the mongooses introduced into the West Indies in the hope that they would exterminate the rats which were ruining the sugar industry did kill some of the pests, but went on to kill off much of the native fauna – mammals, birds and reptiles.

Native species as well as those introduced can become pests to agriculture when man alters the ecology of the environment and provides great quantities of attractive food in the form of standing crops. In the last few decades official afforestation in England has provided habitats for the deer which, whether native or introduced centuries ago, are now derived from feral escapers from captivity. They have flourished and increased so much that their numbers must be kept in check to protect crops in field and garden – the damage a few deer can do in one night to a vegetable patch is heartbreaking. In subsaharan Africa baboons wreak havoc on standing crops, especially on mealies, but are very difficult to keep under control for they are most wary of traps, poison and men with firearms. In East Africa elephants can be such destructive crop-raiders that several thousand are shot every year by the game departments to reduce numbers and drive the survivors back into the animal parks. In Queensland and New South Wales fruit farmers are plagued by flocks of flying foxes, large fruit-eating bats, that spoil their crops, often flying in to the orchards from distant roosts where thousands sleep by day.

In addition to the pests that do extensive damage, other less destructive species have adapted themselves to the man-made environment, and even benefited from it. They have come to terms with their human neighbours, and though often a nuisance they seldom become a serious threat to man's interests. In England foxes live in the suburbs of some towns and forage among the garbage cans at night; it is said that in the dockland of east London there are more foxes to the acre than anywhere in the country. The badger, too, is a frequent near neighbour to human populations that are unaware of its presence. The huge populations of insectivores generally live in peaceful co-existence with man, taking full advantage of the environment provided by him. The only exception is the mole, which if numerous may hamper the use of agricultural machinery by throwing up great numbers of mole-hills. It is surprising what a mess a single mole can make in a garden; to look at the damage one might think that a score of the animals had been at work, whereas the nuisance stops at once when the culprit is trapped. The small rodents of the fields and woods are usually disregarded, though when populations of voles or field mice reach a large size the damage they do is considerable. Such plagues of voles and mice represent the peak of a cyclic population growth, which is invariably followed by a crash in which few survivors remain to begin the next cycle. In Great Britain the water vole causes some damage by burrowing in the banks of rivers and canals, but it is not a major pest. In continental Europe, on the other hand, where it is not confined to the neighbourhood of water, it is sometimes a serious pest to standing crops. On the whole, most of the wild mammals remaining in settled countries are neutrals that cause man little inconvenience; a few species only have become pests.

An aspect of the relationship between man and wild mammals that presents a real threat to his wellbeing is shown by various diseases that can be passed from wild animals to man or to his domestic animals. Some are caused by viruses which, like psittacosis and foot-and-mouth disease, already mentioned, can be passed direct or spread by wild mammals and birds. Among the most dangerous is yellow fever, which is transmitted by mosquitos from man to man. In tropical Africa, and perhaps elsewhere, the mosquito also feeds on the blood of monkeys so that the virus can pass from mosquito to mosquito by way of monkey or man. Virus-B, which can pass directly from monkeys to man without an insect vector, causes a usually fatal disease in him; fortunately it is rare – its victims have generally been research scientists or their laboratory assistants. In Malaysia the virus of scrub typhus occurs in the blood of various species of rat and causes an often fatal disease in man if he is bitten by ticks that have fed on the blood of infected rats.

Attempts to eradicate diseases caused by minute protozoa which are carried from one victim to another by insect vectors sometimes have drastic effects on wildlife. The reduction of human malaria as a world-wide scourge by destroying the vector mosquitos with modern insecticides has, however, had little adverse effect on wild mammals. On the other hand the control of sleeping sickness in man and of trypanosomiasis, the equivalent disease in cattle, has gravely affected the mammal fauna of parts of Africa. Efforts to reduce the diseases have been made in two directions, first to eradicate the tsetse fly, which transmits the diseases, and secondly to eradicate the wild mammal reservoirs of them. Neither method has been fully successful; and chemical injections to kill the parasites in the blood of the victims, though useful, have not yet been perfected. Enormous numbers of game animals have been killed in the attempt to eliminate the wild reservoir of the trypanosome, but in spite of the slaughter, especially in Rhodesia, the diseases remain – indeed it seems impossible that this method can succeed because it is now known that many small mammals, as well as the conspicuous antelopes, harbour the parasites. Even if all the game were killed a reservoir of small mammals impossible to exterminate would remain. Paradoxically the tsetse fly has in the past been the greatest conserver of game animals in parts of Africa by making large tracts of country in the sleeping sickness areas uninhabitable by man and his domestic animals.

In Europe an epidemic of rabies has been spreading westwards during the last decade, spread mainly by foxes, though it probably came originally from wolves in the east. Authority fears that one of the irresponsible fools who smuggle pet dogs into England to avoid quarantine may bring in one infected with rabies, and that the disease may be transmitted to wild foxes. If that happened it would probably advocate a policy of exterminating all our wild carnivores, and muzzling all dogs. We may note, however, that rabies was formerly endemic and was stamped out merely by muzzling every dog for a few years – the disease is transmitted by a rabid animal biting a healthy one – and that no attempt was made to kill wild animals for the purpose. The disease disappeared, and for over half a century it has been kept out by strict quarantine of imported dogs and cats.

Rinderpest provides an example of a disease transmitted to wild mammals by man through his domestic animals. It is a disease highly fatal to ruminants, and was introduced into Africa in imported cattle from which an epidemic arose that swept through the continent in the course of a few years, causing immense loss to stock owners. It also infected the wild ruminants so that hundreds of thousands of game animals perished towards the end of the nineteenth century. Fortunately for the game the epidemic occurred before human pressure on the environment had left no room for wild animals, so that in the course of little more than a decade their numbers had somewhat recovered. Animal populations devastated by natural or man-made disasters usually recover if the animals are left in their natural habitat free from human interference, whether they be seals, whales, white rhinoceroses, polar bears or buffaloes.

7 THE INSECTS

Natura in minimis maxime miranda –
Nature in its smallest creations is most to be marvelled at
Linnaeus

There are more different kinds of insects than of all the other animals together; and there is a greater tonnage of living matter in the shape of insects than in the rest of the animal kingdom. Insects of some kind are everywhere and man cannot escape daily encounter with them directly or indirectly, except perhaps for a few people living in polar regions or air-conditioned apartments. Most members of this vast host are neither harmful nor useful to man; a large number, though but a small fraction of the whole, is destructive to his interests, and a few are directly valuable.

Until recent years the neutrals, the majority of insect species, were little affected by man but, though they appear to be of little importance to him, he is now becoming increasingly important to them by reason of the changes he is bringing about in the environment. The great quantities of chemical insecticides distributed throughout the world to protect crops from damage by insects, and often used with careless indiscrimination in amounts far larger than necessary, have slopped over from their legitimate fields to pollute the environment with poisons even to the uttermost ends of the earth. The poisons enter the food chain not only by way of the insects but through many other creatures which become contaminated with them indirectly. Dieldrin, aldrin, DDT and other chlorinated hydrocarbons and their derivatives can be found everywhere in the tissues of animals, from the eggs of golden eagles in Scotland to the fat of penguins and seals in the Antarctic; no place is free from them throughout the world – is it a cynical prophetic joke that one of the first men to set foot on the moon was Aldrin?

In Great Britain the conspicuous decrease in the numbers of butterflies, dragonflies and other insects during the last few decades has been attributed to the widespread indiscriminate use of insecticides. This is doubtless true to some extent, but the truth of the allegation has not been proved, nor is it probable that this is the sole cause of their comparative scarcity. Many other causes may have contributed, such as the spread of built-up areas and the changing methods of agriculture, through which the former balance of the ecology is being altered and many habitats are being destroyed. The universal use of herbicides in agriculture to kill the weeds of cultivated crops must also have had a profound effect by destroying the food plants of many species. There is however often a natural cycle in the size of insect populations, independent of man's activities, and due to climatic and other environmental changes.

The spread of the white-admiral butterfly throughout the west of England and into Wales during the 1920s and 1930s from its headquarters in the New

Forest, so that it became a common species in places where it had never appeared before, is a good example of a population increase apparently independent of any activity of man. The black-veined white butterfly became extinct in England, and the large tortoiseshell butterfly became scarce, long before modern insecticides were invented, so that human influence appears to have had no part in their decline. On the other hand the extinction of the British sub-species of the large copper butterfly was the direct result of the destruction of its habitat with the draining of the fens – when it was reduced to a small remnant its rarity and commercial value as specimens led the collectors to finish it off. There has been little noticeable change in the abundance or scarcity of insects since the use of some of the chlorinated hydrocarbons was stopped in Great Britain a few years ago; perhaps we are in the trough of a natural cycle, and a peak may yet follow it.

Most of the insects against which insecticides are used are pests of cultivated crops; no one would regret the extermination of the pests if it could be done without also killing the harmless and the beneficial species. Yet the pest problem is of man's own making; the attempt to grow pure stands of crop plants is itself a gross interference with the ecology of wildlife, and the vast increase in the populations of insects that feed upon them is thereby invited to reach pest proportions. The efficiency of modern insecticides under such conditions has enormously increased the world production of food, so that larger human populations can exist on the verge of starvation – the larger cake has to be divided among a larger number of mouths in a vicious spiral of growing numbers and production without growing prosperity.

The locust is an example of an insect that probably swarmed to pest proportions long before man had any crops to be devastated. When crops were cultivated they were despoiled by locusts just as was wild vegetation, for the insects eat any green plants and do not selectively attack crops. The arrival of swarms of migratory locuts, particularly in Africa, is a disaster for farmers and the countryside in general, because the vegetation is completely destroyed wherever a swarm settles to feed. Many rule-of-thumb methods for combating locust swarms were tried with little success until the problem was tackled scientifically about fifty years ago.

Research showed that locusts do not always live in swarms; in their native habitats they live as scattered individuals like any other grasshoppers, but when seasons are favourable the population numbers build up and the insects begin to congregate together until a large swarm is formed, which flies away seeking new feeding grounds. The locusts of the solitary phase differ in details of colour and structure from those of the swarming or gregarious phase. The swarms that have flown away from their permanent home may travel hundreds of miles eating everything green on which they settle, the females laying batches of eggs periodically along the route. The eggs hatch to release swarms of young locusts known as hoppers, for their wings do not grow until they become adult after many moults of the skin. The swarms travel on for several generations but in the end they die out, though they are later followed by new ones derived from the solitary phase in the permanent home.

The international locust control organisation co-ordinates the work of locust research laboratories in many lands; constant watch is kept for the appearance of swarming hoppers so that control measures may be taken before the swarms get on the wing. The problem of locust control is complicated because there are several species that swarm, with permanent habitats in different places. The dessert locust, for example, comes from a permanent home in Asia Minor whence its swarms may travel thousands of miles into central Africa, breeding and producing new swarms on the way. Disastrous as the swarms are to man they are no doubt very welcome to the many kinds of animal that feed upon them – in Africa the birds such as storks, pratincoles, wattled starlings and others are known collectively as "locust birds" because they follow the swarms to feast upon the insects. Locusts are evidently highly palatable, for many kinds of birds, reptiles, and mammals in captivity eagerly accept them when offered as food.

Although natural predators have little noticeable effect on the numbers of locusts when swarming, they do keep the numbers of many other insects destructive to plants within reasonable bounds – how often does one find wild plants covered with dense masses of aphids like those which attack our cultivated roses and other crops? The insectivorous birds, such as warblers and tits, and predatory insects such as the ladybird beetle and its larva, generally preserve an ecological balance in the wild, and it is mainly our tender cultivars that fall victims to the pests. Although we now use vast quantities of chemical insecticides in our efforts to eradicate such pests, natural predators have sometimes been successfully used for the purpose.

When the cottony-cushion scale insect, one of the plant bugs, was accidentally introduced from Australia to South Africa and California it infested the citrus orchards and threatened ruin to the growers. A natural predator on the bug, the cardinal ladybird, was then brought from Australia and released in the orchards where it soon reduced the infestations to negligible size – this was one of the early successes in biological control. Minute chalcid wasps have similarly been used to control the black scale-insect which is destructive to citrus fruits; and other species of the same family of wasps have been used to control wooly aphis and codlin moth, which are serious pests in apple orchards. Another method of biological control is to break a link in the chain of events that make up the life cycle of the pest, as in parts of Africa where the bush along water courses has been destroyed in an attempt to deprive the tsetse fly of its breeding habitat, though with only moderate success.

A similar scheme was proposed during the 1939-45 war by a British official with more enthusiasm than knowledge, when he learnt that the black fly, an aphis destructive to field and broad beans, lives over winter on spindle trees. He actually obtained an allocation of many thousands of pounds to pay for uprooting and burning all the spindle trees in the country. He did not know that spindle not only makes a medium-sized tree, but is found in hedgerows which are full of bushy spindle, and in woodlands which are scattered with spindle saplings: eradication of the plant would have been impossible. Before the money was made over someone pointed out that black fly are also air-borne into the country from the Continent every year, so that removing the winter host plant would not eradicate the pest. The money was saved – but the official was not dismissed.

Insects have been used with success not only in the biological control of other insects, but for eradicating harmful weeds. During the last century, as noted in Chapter 2, the prickly pear cactus was introduced from America into Australia and South Africa where it flourished exceedingly and covered thousands of acres of potentially valuable agricultural lnd. A cochineal insect, one of the plant bugs, and the prickly-pear moth, both of which feed on the prickly pear in its native land, were introduced into Australia in an attempt to eradicate the plant. Although both insects feed on the cactus in America, they are in some sort of ecological balance with it so that though they may damage they do not destroy their food plant. In their new environment in Australia they dramatically reduced the weed which had flourished so exuberantly under the same new conditons, but when these species were introduced into South Africa they were useful in reducing the nuisance but not so completely successful in eradicating it.

A new method of biological control, also noted in Chapter 2, has recently been developed in the hope of eliminating the tsetse fly from parts of Africa. It consists in breeding large numbers of the flies in captivity and sterilising the males by subjecting them to a dose of X-rays sufficient to make them infertile but not to kill them. The sterilised males are then released in fly areas where-upon they mate with wild females, which are thus unable to produce live off-spring. The method has produced a reduction but not the extermination of local populations. A similar attack on the mosquito that carries yellow fever has been proposed.

Biological control by the introduction of predators to destroy unwanted animals or plants has thus sometimes been successful, but it is a method that must be used with great discretion, for it can prove to be a two-edged sword and have disastrous unforeseen side-effects. Like the mongooses introduced into the West Indies to kill rats, the foxes introduced into Australia to kill rabbits failed to complete their intended task and multiplied to become a nuisance and a threat to the survival of the native fauna. With such examples in mind the Australian agricultural authorities were careful to make sure that the rust fungus they introduced to destroy the skeleton-weed would attack that plant and no other. The weed is a serious pest in south-east Australia but the fungus causes a fatal epidemic infection greatly to the benefit of cultivation. It is notable that in many of the instances of biological control, whether successful or not, the pest attacked has been introduced into a new environment either intentionally or by accident, so that the problem to be solved is one that man has set for himself.

The insect pests of cultivated crops are far too many for more examples to be mentioned here; every species of crop plant has at least one species of insect that battens upon it, and most of them have many. When the insects have been defeated in the field a different horde of pests descends upon the harvest safely gathered in, as soon as it goes into storage. The damage done by the pests of stored products can be so serious that its control is a never-ending war. Nothing is free from attack, from cereals as grain or manufactured products, natural fibres and textiles made from them, paper and books, timber in houses or furniture made from it, to preserved meats such as bacon, biltong and stockfish, and such improbable things as dried tobacco, pepper and other drugs more poisonous to man. Most of these pests when living in the wild are scavengers of dead or moribund animal or vegetable matter, though some of the wood-boring insects attack sound standing timber. It is only when they get into the large quantities of their food amassed in man's warehouses and dwellings that they multiply exceedingly and destroy the fruits of his toil. They are difficult to eradicate because the means used must not contaminate foods with poison, nor be so drastic as to inflict damage worse than that done by the pests upon other goods. Fumigation with volatile poisons that evaporate completely, such as carbon dioxide, can be used to destroy pests in foodstuffs, and the chlorinated hydrocarbons, especially DDT, have enormously reduced the damage done by insects to materials and manufactured goods other than foods.

Man has, however, found a way of using one species for his advantage: those who prepare small animal skeletons for museums let the larvae of bacon beetles – dermestids – eat away the remains of dried flesh attached to roughly cleaned skeletons and, by removing the bones from the insects at the right stage of the meal, obtain beautiful preparations of the most fragile vertebrates with all the bones attached to each other in their natural positions by the remnants of the ligaments that held them together in life. The work must be done away from collections, for the accidental introduction of the beetles into a museum can wreak havoc among the stuffed animals – as does another species that manages to work its way into the most tightly closed store boxes, so that it has gained for itself the name of museum-beetle.

The pests of crops and stored products are an instance where man, far from damaging wildlife by his activities, has caused an enormous increase in its abundance much to his discomfort. Man has accidentally introduced the pests of stored products to all parts of the world, and, in spite of all his efforts to exterminate them, including the use of modern insecticides and biological control, they are still with us. Only one out of all the host has endeared itself to man – the cricket on the hearth, which lived in crevices near the fireplace of old-fashioned kitchens, coming out at dark to feed upon crumbs and other debris. The chirping of the males was to some people a cheerful song, though to others it was a maddening nuisance. Crickets find no place in modern kitchens lined with shiny tiles – they went out with kitchen ranges and domestic bread-ovens. Crickets, together with cockroaches, have been banished by DDT from such old-fashioned bakehouses and similar places as may remain, but often swarm on town rubbish dumps where decaying organic matter provides them with warmth and food. They might there be regarded as beneficial insects because they remove offensive matter, though their presence is usually looked upon with disfavour by unimaginative authority.

All the insect pests of stored products, and many pests of growing crops, inflict their damage by eating the fruits of man's labours, nibbling away with their horny jaws like the proverbial weevil in a nut. They may truly be called biting insects, though the term is incorrectly applied to the blood-sucking and plant-juice sucking insects which do not bite with horny jaws. Their mouth parts are formed so that they pierce the skin of their victims as with a hollow needle through which they obtain their liquid food – when we provide the meal we commonly but inaccurately call the resulting lesions insect bites or stings – the French *piqûre* is a more just term.

The 'bites' of blood-sucking insects cause an inflammatory reaction which is often intensely irritating, and if the aggressive insects swarm in myriads, as do the midges familiar to visitors to northern Scotland, their attacks are so intolerable that man is compelled to flee. The torture inflicted by blood-sucking insects is not, however, the reason for their great importance; far more sinister and dangerous is the fact that their hypodermic method of feeding can transfer the germs of disease, whether bacteria, viruses, protozoa or even microscopic worms, from man to man, beast to beast, or from one to the other in either direction.

Some of the insects merely carry the infection from one person or animal to another mechanically, but others are an essential link in the life cycle of the disease-producing organism, which cannot be transmitted directly because it msut go through some of its development stages in the insect. Malaria is probably the most widely spread disease transmitted in this way. The disease is caused by a minute parasitic protozoon that lives in the blood corpuscles of man and other vertebrates, releasing toxins into the blood when the corpuscles break down at the end of each cycle of the parasites' asexual reproduction. Each release of toxin brings on a bout of fever, the length of the cycle according to the species of parasite, causing a tertian or quartan ague. If blood containing the parasites is sucked by one of several kinds of mosquito they escape digestion, enter the mosquito's tissues, reproduce sexually to give numerous progeny which enter the mosquitos' salivary glands, whence they get into a new vertebrate host when the mosquito feeds. The details of the parasite's life history are complex, as indeed is man's relationship with the two forms of wildlife: the mosquito depends upon man for its food, the malarial parasite depends on both man and mosquito, but man gains nothing from the transaction.

Malaria has brought suffering and death to millions of people, and has even influenced the course of human evolution in some parts of the world. In parts of Africa some people suffer from sickle-cell anaemia, in which the red blood cells shrivel to a characteristic shape. The condition is determined genetically; people heterozygous for the sickle-cell gene have a high resistance to malarial infection but do not suffer from anaemia, though those who are homozygous languish under the often fatal anaemia. The malarial parasite has thus acted as the agent which favours the survival of people carrying the gene, though at the cost of death to those born homozygous for it. The malaria parasite has also set man to work on vast drainage schemes to deny breeding places to the mosquito, but it was not until the invention of DDT to kill the insects, and of powerful anti-malaria drugs to kill the parasites in the blood, that control of malaria on a wide scale was achieved – and produced a problem of human population increase that is still unsolved.

Trypanosomes are also protozoal parasites in the blood of vertebrates – the species transmitted by tsetse flies and causing sleeping-sickness in man and nagana in domestic animals have already been mentioned. Although the fly can transmit the trypanosomes directly if it bites a second victim shortly after feeding on an infected one, the trypanosomes need to live in the body of the fly for about three weeks to complete their development and life cycle. Wild animals of several sorts are the reservoirs of trypanosomes; they are carriers of the parasites which do not produce symptoms of disease in them. Another kind of trypanosome, which causes Chaga's disease of man in tropical South America, is transmitted by a large bug, the barbeiro, in which the parasite undergoes part of its life cycle. Armadillos are the wild hosts of the parasite and form a reservoir of carriers which show no symptoms of the disease. Many other species of trypanosome, and of other kinds of protozoa, similarly pass part of their life cycle in the body of the insect vector which transmits them from host to host, and produce serious diseases in man and his domestic animals; they are so numerous that the above examples must suffice.

Parasites much more complex in their structure than protozoa also pass part of their life cycle in the bodies of insect vectors, and produce diseases when transferred to susceptible hosts. The larvae of many species of Filaria and allied blood-worms, minute nematode worms that live in the blood of vertebrates, have to undergo part of their development in the insects that transmit them. The different diseases that they produce are known collectively as filariasis – in one the worms live in the blood vessels deep in the body of the vertebrate host by day but come to the skin capillaries at night when the mosquitos that transmit them feed. In another they block the lymphatic vessels and produce a crippling elephantiasis in the victim. The West African disease loa-loa is caused by similar worms transmitted by a tabanid fly, *Chrysops;* but they appear in the skin capillaries by day rather than by night, as the fly feeds during daylight. Blackflies, *Simulium damnosum,* transmit the worms that cause the serious equatorial African disease, 'river blindness'.

Some tape-worms are carried from one definitive host to another by insect vectors in which they pass the intermediate cysticercus stage of their lives. One of the dog tape-worms is thus carried by dog fleas; the flea larvae, feeding on the debris at the bottom of the kennel, swallow the worm eggs which hatch to form cysticerci in their bodies where they remain encysted through the larval, pupal and imago stages of the flea. When the dog, nuzzling and nibbling at his flea-bites accidentally swallows a flea, the cysticerci are released into its intestine where they grow into adult worms. A tape worm of the rabbit similarly makes use of a mite as its secondary host.

Apart from these complex life cycles in which disease-producing parasites divide their stages of development between different hosts, there are many others which are merely transferred mechanically from host to host by insect vectors which act as inoculators or living hypodermic syringes. Some of man's most terrible diseases are spread in this way. The virus of yellow fever, a disease which has often been epidemic in warmer parts of the world, is spread by a mosquito which carries the virus from the sick to the healthy. The disease brings horror and panic to stricken populations, not only through the vast numbers it kills, but through the swiftness of its action – people apparently healthy in the morning may be dead before nightfall. Apart from prophylactic injections the disease has been controlled by destroying the breeding places of the mosquito, a tremendous task because it breeds in all sorts of small puddles and pools, even in empty sardine tins thrown on a rubbish heap. The mosquito is widespread throughout the warmer parts of the world, and now great care must be taken to ensure that infected mosquitos are not accidentally carried from one country to another in aeroplanes.

Bubonic plague, the black death of the Middle Ages and the plague from which Mr. Pepys fled with his coach-load of guineas in the year before the great fire of Londin, is caused by a bacillus inoculated into the blood by fleas. Gerbils, rodents resembling large mice and found throughout Africa and southern Asia, are the natural reservoir of the bacillus which is passed from one animal to another by their fleas. If an infected gerbil flea feeds on a rat, particularly one of the forms of the black rat, the rat becomes infected but, unlike the gerbil, develops symptoms and generally dies of the plague. If, further, an infected rat flea feeds on man the infection is passed to him. Bubonic plague is thus spread through human populations by fleas from infected rats, but if the bacillus infects the human lung the disease takes on the pneumonic form that can be passed from man to man direct. As rats have been taken unintentionally to all parts of the world by man the native populations of wild rodents have in some places become infected with plague and act as reservoirs in South Africa, California and other parts of the Americas. There appears to be no transmission of plague by fleas from gerbils to man so the disease can exist undetected for years until it happens to be passed to rats and thence to man. When the rat flea feeds on infected blood the bacilli accumulate in its gut and form a mass completely blocking it, so that the insect cannot digest its blood meal. The hungry flea ravenously sucks blood in trying to feed, but the blockage forces it to regurgitate, thereby passing some of the bacilli into the blood of its victim. The flea dies of starvation, the human victim dies of the plague. The disease is controlled by keeping rat populations as low as possible and particularly by preventing the accidental importation of rats from places where plague is endemic.

In most countries there are strict regulations for the 'de-ratting' of ships, and the matter is so serious that in British ports the deadly poisons fluoroacetamide and sodium fluoro-acetate are used, poisons so deadly that their use except in ships was prohibited after several horrid accidents. All the fluorine in the world is safely locked up by nature in harmless and even beautiful inorganic compounds such as Iceland spar; it is only when man interferes and puts it into organic compounds such as these and polytetra-fluoretheylene – the amazing frictionless plastic PTFE – that its destructive powers are revealed in substances never seen before. The accidentally man-made reservoirs of plague in the native populations of wild rodents are beyond the possibility of control, and will remain as a potential source of danger, perhaps permanently.

Fleas as transmitters of another disease suddenly became of great importance to man in the early 1950s when myxomatosis appeared in the vast populations of rabbits, then his biggest agricultural enemy in Europe. Myxomatosis is endemic in South American species of rabbits, in which it does not produce acute symptoms, but when the virus was inoculated into European rabbits it brought an epidemic with nearly a 99 per cent kill. The virus had been released some years before into the rabbits of Australia and had brought immense relief from a pestilential alien introduction. In that country the virus is transmitted by a mosquito, but in Europe the vector is the rabbit flea – as the virus is transmitted mechanically almost any blood-sucking insect could presumably be a vector. In Great Britain there were proposals for exterminating the remnant of the rabbits that the disease had missed, a difficult operation which, however, could have succeeded. But sentiment clouded the arguments and a great opportunity of ridding the land permanently of a pest introduced some eight or nine centuries before was lost – it was even made illegal to spread the disease deliberately.

Myxomatosis thus became endemic; and strains of reduced virulence appeared as did rabbits with increased resistance. The surviving rabbit population now goes through local cycles of abundance and decline as numbers build up until a virulent strain of virus infects them. Myxomatosis was deliberately introduced into Australia in the hope of defeating the rabbit plague – it caused a spectacular decrease in the number of rabbits but did not exterminate them. It was introduced in northern France with the intention of clearing a small estate; it succeeded beyond expectation and rapidly spread through most of Europe. How it came to England is not known officially; no-one has admitted introducing it, though an unsuccessful attempt was made in the 1930s to introduce it to the island of Skokholm off the Pembrokeshire coast. It failed, apparently because there were no rabbit fleas on the island to act as vectors, a state of affairs very different from that on the nearby island of Skomer where the numerous rabbits carried heavy infestations of fleas.

Lice are blood-suckers that transmit diseases to man, though not by injecting him when they feed. If a louse takes blood from a person suffering from certain virus diseases such as typhus, trench fever, relapsing fever, or others, the virus passes through the insect's gut and is present in the faeces deposited on the victim's skin. The irritation of the puncture makes the person scratch the place thereby rubbing the virus in the faeces into the scarification – he unknowingly inoculates himself. The virus is also spread on the skin when a person squashes a louse and thus smears the contents of its gut over the skin.

Although wingless and minute, body lice pass readily from person to person by casual contact. They breed rapidly, especially on those who wear their clothes for long periods without taking them off, such as soldiers in the trenches of the First World War, or down-and-outs who sleep rough. The subspecifically different head louse is common on children, especially in industrial towns, and is said to have increased in numbers recently owing to the fashion for long unkempt hair among adolescents. The crab louse, which generally lives only among the pubic hair, is not known to be a transmitter of disease – it has been called more disreputable than dangerous. The invention of DDT enabled whole populations of people to be quickly freed from lice and the possibility of typhus epidemics, for the chemical is very fatal to the insects. Before that, the application of grey mercurial ointment was the usual, but less efficient, remedy to get rid of lice – the grey paint used on ships of the Royal Navy was jokingly known as "crab fat" on the lower deck.

Although many kinds of flies are blood-suckers, and thus carriers of disease to man, the common house fly and its relatives have a proboscis for sucking up liquids but not for puncturing the skin. Flies when feeding moisten the food with saliva, and regurgitate the contents of the stomach, to make a solution that can be sucked back. As they feed on all manner of decaying filth as well as on man's food if it is left exposed, it is not surprising that they contaminate the one from the other with the bacteria in their guts and sticking to the hairs of their bodies and feet. In this way they cause out-breaks of gastro-intestinal infections from "food-poisoning" to typhoid. Flies breed rapidly during warm weather, laying their eggs in moist decaying organic matter on which the larvae feed; each female lays five or six batches of about 150 eggs, and the life cycle from egg to adult fly can be as short as ten days, so the potential number of food contaminators is great. Until about the end of the first quarter of the present century fly contamination was responsible for the annual out-breaks of summer diarrhoea, a disease highly fatal to infants in crowded industrial towns. When horses were replaced by automobiles, and stables were turned into garages, the manure pits and heaps in which flies bred disappeared so that the fly population was greatly reduced and the disease has been practically eliminated by improved hygiene.

Several species of fly beside the common house fly come into houses, and all are dangerous contaminators and spoilers of food. A rather small fly, the lesser house fly, is the kind that used to fly in small flocks jerkily and erratically circling chandeliers or electric lights pendent from ceilings. Flies are now much less plentiful in houses than they were a few decades ago; their numbers have decreased with those of town horses, and with the use of fly-killing sprays, aerosols, and nerve gases slowly liberated from impregnated plastic strips. The fly nuisance is today mainly a memory of older people in Great Britain, though in other parts of the world flies are still a plague bringing disease, discomfort, and disgust to suffering humanity.

Some of the many species of "biting" flies have already been mentioned, but two English species merit a glance. The stable fly, which looks much like the house fly on casual inspection, has a skin-piercing proboscis, and feeds by sucking blood. It attacks man as well as domestic animals, and shows a strong preference for pricking the skin of the ankles. The English summer pastoral scene, popular with the artists of the nineteenth century, showing water-meadows with cattle standing knee-deep in a placid, slowly-flowing river as they ruminate, is not so peaceful as they thought, for the cattle are in the water to avoid the attacks not only of this fly but of another, the warble fly, which is much more damaging.

The warble fly is not a blood-sucker but lays its eggs on the hair above the feet of cattle whence the larvae which rapidly hatch out are transferred to the mouth when the animals lick themselves. The larvae at first attach themselves to the lining of the gullet and later burrow into the tissues, finally coming to rest beneath the skin of the back, where they form an abscess with a hole for breathing. They feed on the body-fluids of the host, and when fully grown worm their way out of the hole and drop to the ground to pupate. They cause much economic loss because though the skin heals over the place, the scar comes out when the hide is tanned to make leather – a hide with warbles spoiling its best part with numerous holes is of little value. It is peculiar that cattle are terrified and stampede when they hear the characteristic buzz of the warble fly near them, for the fly does not bite them and they cannot know that its attentions will result in painful warbles many weeks later.

The larvae, known as bots, of a related species live in the stomach of horses which they enter in a similar way, but those of yet other species that live in the nasal sinues of sheep and deer hatch from eggs laid in the nostrils of the victims. A common blood-sucking fly in England, which is annoying rather than dangerous, is the clegg or horse-fly, which silently and gently alights on the skin and manages to insert its probosics so that one only becomes aware of it when it starts feeding, producing a stinging feeling and leaving an irritating swelling. In the western counties cleggs are called "old maids" – because they have such sharp tongues. Many related species of gad-flies in other parts of the world are painful blood-suckers, sometimes making their habitats almost uninhabitable by man.

The larvae of flies that feed on the living flesh of mammals are perhaps the most repulsive to human feelings. Those of the greenbottle fly "strike" sheep and cause serious losses unless the sheep are regularly dipped in solutions poisonous to the fly. Many species in other countries similarly attack various mammals and birds – the larvae of the berney fly of the tropics even parasitise man, living in subcutaneous cysts usually on the neck. In Europe the larvae of bluebottles or blow flies, which usually feed on carrion, are sometimes found in open wounds of man and other mammals where, though they appear revolting, their excretion of urea and allantoin actually help to stimulate healing.

There are, however, some insects that are more beneficial to man than the larvae of blow flies; the value of their products and services, though comparatively small is something to put in the balance to counterpoise the immense damage their relatives inflict on humanity. Few species of insects are or have been important as articles of food; the witchity grub of the Australian aborigines, and termites in parts of Africa are well known exceptions. The ancient Romans are said to have relished the larva of the goat moth as a delicacy, and even today fried agave worms – the larvae of a moth – are pushed commercially as fashionable delicacies, and some people find Japanese ants in chocolate sufficiently desirable to pay high prices for them.

It is, however, the substances produced by insects rather than the creatures themselves that are valued as useful food. The manna of the Bible is the honey-like secretion of a scale insect that infests tamarisks, and is still sometimes eaten by man. By far the most important product, known and used by man from earliest time to the present, is the honey made by several kinds of bee, social species of the aculeate hymenoptera. The four species of honey bee, three natives of tropical Asia and one of Europe, are the most productive, and the two that nest in hollow trees and rock crevices are the most useful. The frequent difficulty of taking the honey from barely accessible hollows led man to provide suitable places for nesting, as he still does in Africa where he hangs hollow logs from the branches of trees, and gathers a honey crop when they are filled. It is but a step from providing such primitive housing to the development of the more elaborate modern hives. Honey bees, however, are not domesticated, and although skilful apiarists can manipulate them to their own advantage and have selected good strains for their purposes, the insects remain as wild as those that do not take advantage of the artificial nesting places provided.

Bee-keeping was formerly of much greater importance than at present because honey was the only form of sugar available to man until the wide cultivation of sugar cane began in the sixteenth century, and that of sugar beet in the nineteenth. Now, after millennia of being robbed by man, the honey bees are getting their revenge, at least in South America. European honey bees had long ago been introduced by apiarists, from whom some had escaped so that the species is well established in the wild. Escaped swarms from a recent importation of particularly aggressive bees have established themselves in the wild, increased and interbred with the long-acclimatised, more placid bees, to produce equally aggressive hybrids. These bees are so fierce that in some parts of Brazil cattle, horses and even men have been killed by their stinging attacks – they are spreading rapidly and threaten to make some regions almost uninhabitable. If the Brazilians try to exterminate the killers by the indiscriminate use of insecticides, as they probably will, they will do immense damage to their spectacular insect fauna, with consequent ecological damage too great to be predicted.

Insect products valuable to man, though of less importance than honey, are provided by various plant bugs, mainly scale insects. Lac, extensively used as shellac until the invention of synthetic polishes, is the secretion of an Asian scale insect, as are the Chinese and Indian white waxes which used to be more important in commerce than they are now. The old world plant-bugs that produced a red dye were superseded commercially by the cochineal insect of the New World, whence it was introduced into the Canary Islands and the Mediterranean countries.

Bees, and other insects that visit flowers to gather or feed on nectar or pollen, are valuable to man by pollinating his fruit and some other crops, but in the course of evolution they have played a much larger role. The parallel and mutually dependent evolution of flowering plants and flower-visiting insects has determined the entire character of the environment for man as for all other wildlife. Had this extraordinary inter-relationship between insects and plants not evolved, the vegetation of the world would have been completely different from that we see, and the evolution of animal life would probably have been guided into paths other than those it has followed.

Only one insect has been domesticated, the silkworm moth, which is entirely dependent upon man and does not exist in the wild, unlike the form from which it was probably derived. The long and fascinating history of the silk industry, and the influence it has had upon the spread of civilisation is, however, rather outside our subject of the interaction of man and wildlife, though wild silk-producers of other kinds are still exploited in Asia. These are the large moths, reared in captivity, from the cocoons of which the rather coarse Tussor silk is obtained.

Although some insects are directly useful to man, and many are looked upon as beneficial because they prey upon others that are noxious to him, man's attitude to insects is in general one of perpetual warfare against an immense horde hostile to his every interest. Yet, as already mentioned, this horde is but a small part of the total insect fauna most of which is neutral towards him. From their numbers and almost infinite variety it is a war he is unlikely to win; at most he can hope to contain the enemy and ward off the worst attacks. This is just as well, because insects enter into every detail of the ecology down to the smallest, and are essential elements in the economy of the environment – among many other things, a large portion of the world's fauna depends solely upon insects for food. Insects also play a large part in the recycling of organic matter, often carrying out the first steps of the process which is continued by fungi, protozoa and bacteria. Where man has succeeded in gaining the upper hand over the insect vectors of disease the resulting unwanted human population explosion shows that the biological control of his numbers mediated through insects was an advantage to him as a species, however horrible it may have been to the individual. In the broad view the lines between harmful, harmless and beneficial insects are hard to draw, for sometimes out of evil comes forth good. It seems to be inevitable that for every problem set by wildlife that man solves he sets himself a greater.

8 THE OTHER INVERTEBRATES –

Men have died from time to time, and
worms have eaten them, but not for
love. – *As You Like It.*

About a million different kinds of animal have been given names by naturalists, but many more remain to be found and recognised, especially among the inconspicuous, the small and the microscopic. The insects, to which Chapter 7 was devoted, make up the major part of this assemblage, running to about 700,000 different species; whereas the rest of the invertebrates number less than half as many, with about 257,000 species. Of these the molluscs are the largest group with about 100,000 species; the arachnids – spiders, scorpions, mites and ticks – come next with about 30,000 species; followed by the crustacea with some 25,000. The remaining 102,000 species of invertebrates are distributed between numerous groups or phyla, some large and some small, including the protozoa, the echinoderms, coelenterates, and the several groups of animals without close zoological affinitiies included in the vernacular term "worms". In contrast to these huge numbers it is interesting to note that all the vertebrates come to only 42,600 species together, of which about 23,000 are fishes and only 4,000 are mammals.

Judged by the number of species and the immense variety of habitats exploited all over the world, the insects are the dominant group of animals. It is, perhaps, fortunate that the structure of their bodies is such as to preclude any of them reaching a large size – the biggest is barely larger than a mouse in bulk, the longest not more than a foot in length and thin at that, and of those that fly the greatest wing-span is likewise less than a foot and the body comparatively small. The limitation of size is due to the way that insects breathe; air is taken to all parts of the body by a system of branched tubes, the tracheae, which ramify in progressively smaller divisions among the various organs. The tracheae open on the surface of the body through small pores, the spiracles or stigmata, which range in number from two to ten according to kind. This system of tracheal breathing is highly efficient in animals of small size, but there is a point in the slope of increasing size beyond which it breaks down and is unable to supply the oxygen demand of the tissues. In addition the exoskeleton, the outside hard chitinous covering of the body and limbs, necessarily becomes thick and heavy with increasing size, thus giving the animal too great a proportion of dead-weight to weight of living tissue. Had there not been these limitations the insects might have become the dominant group, not only by their numbers and variety, but also through the size of some of their members – but this train of thought leads to the realms of science fiction.

Taken as a whole, the invertebrates other than insects occupy paractically every imaginable habitat throughout the world and, as many are of minute size, often in immense numbers. Members of one group or another form meshes in the ecological web everywhere, and like other animals play their part in shaping the environment. In that sense all are of equal importance, but in their relationships with man some are of outstanding and immediate importance. Man uses some directly as food, and obtains useful products from others; some are pests of his crops, parasites of his body and of his domestic animals, or vectors of disease; and some are essential links in the food chain leading to his dining table.

Two phyla, the mollusca and crustacea, collectively called shellfish in commerce, provide most of the species that man uses as food. The majority are wild animals that he gathers from their natural habitat, but a few are brought into captivity or provided with an artificial environment in which they are protected from predators and their growth to edible size is facilitated. Both the snail-like gastropods and the mussel-like bivalve lamellibranchs are molluscs that are slow-moving, sedentary or even attached to a substrate, but the squid-like cephalopods are active and very mobile; with the exception of the numerous terrestrial gastropods all live in water, generally salt. In correlation with their limited or lack of ability to move about most of the gastropods and the bivalves are protected by calcareous shells secreted by the outer layer of the soft body within – the soft body which makes them so sought after and relished as food by many vertebrates from fishes to man.

The gastropods are, in general, less palatable and less valued by man, though other animals make no such discrimination. The marine gastropods of Europe, such as whelks and periwinkles, are traditionally the popular food of the lower classes, at least in Great Britain, where they are one of the treats of a visit to the seaside. The large land snail, or Roman snail, on the other hand, is the delight of the gourmet and has been highly prized for several thousand years. It is one of the few that are reared and fattened in captivity, the *escargotières* or *parcs à escargots* of Burgundy being the successors of the *cochlearea* of the ancient Romans. Far from suffering as a species from long-continued human predation the edible snail has become more widespread; it is not native to Great Britain and became a member of the fauna through introduction by the Romans – its shells are not found with those of other species in deposits of pre-Roman times. Other smaller kinds of snails are commonly eaten by the peasant classes of Europe, and in England the garden snail was until fifty years ago consumed in some places not only as a food but as a medicine for chest complaints, as described in Chapter 2.

Of all the European marine gastropods the limpet must surely be the least appetising – it is very tough for one thing – though the coastal dwellers who made the enormous middens of shells in prehistoric times must have either liked it or accepted it as an alternative to starvation. Another gastropod, the ormer of the Channel Islands and Normandy coasts, with the similar abalone of the Pacific coast of America, is also a tough subject which is nevertheless held to be a delicacy when soaked in a solution of soda and then vigorously pounded with a rolling pin. It is so highly esteemed in the Channel Islands that it has long been protected there by a close season; but it has of late been so heavily over-gathered that the threat of extermination recently led the States to decree a total prohibition on taking ormers for a period of five years.

It is, however, the bivalve lamellibranchs that today form the largest part of the molluscan shellfish eaten by people with a good standard of living. They are filter-feeders, drawing a current of water into the shell by the action of countless minute whip-like cilia on the gills to separate planktonic organisms or suspended organic debris and pass them to the mouth. Most gastropods, on the other hand, feed by rasping away plant or animal matter with the tongue-like radula covered with numerous minute teeth.

Unlike the gastropods, most bivalves do not need to move about to find their food, since the water around them is full of it, and consequently they often live as closely crowded neighbours forming beds or scalps. This way of life makes it easy for man to gather such shellfish as clams, cockles, mussels and oysters, and also to cultivate in an artificially improved environment the more desirable sorts, among which the oyster takes first place. The kinds that live in tidal waters, such as clams, cockles, mussels, and sometimes oysters, are gathered at low tide, and others living beyond the tides are taken by dredging in no great depths of water. Human pollution of the environment with sewage, though not with industrial wastes, far from being detrimental to sedentary molluscs provides them with a rich source of food, either directly or through the food chain of unicellular plants. Mussel scalps can thus become so overcrowded that it is necessary to thin the animals out in order to grow a smaller quantity of large ones rather than thousands of useless small ones.

The oyster, however, is the chief species that is artificially cultivated, as the one most highly esteemed gastronomically. Its method of reproduction and mode of life also make is specially suited for culture, as described in Chapter 3. When oysters spawn, the enormous numbers of planktonic larvae settle on a suitable substrate after a short free-swimming period. Much of the art of oyster culture depends upon securing a good supply of young oysters, and to this end the farmers provide frames stacked with pottery tiles covered with a film of lime-wash. The larvae settle upon these and attach themselves by one shell, which is welded onto the substrate; the settling is known as the "spat fall". A good spat fall leaves the tiles crowded with minute oysters, which can be removed without damage as soon as they reach the proper size by scraping the lime coating from the tiles. Thereafter they are grown-on with constant attention in the parks, being transplanted seawards in the estuaries as they grow older.

There is a large trade in young oysters for growing-on and fattening, because in many places that are suitable for oyster culture the spat fall is often scanty, irregular, or a failure. Many English oyster beds are thus stocked from elsewhere when they are not self-replenishing, and oyster parks on the west coast of America are stocked with young oysters from Japan. Oysters cannot be regarded as in any way domesticated; they are truly part of wildlife, yet by modifying the environment man exploits them much to his advantage – and to theirs, until the final moment of truth. On the other hand the necessity for artificially restocking some English oyster fisheries has been brought about by over-dredging unfarmed natural beds of wild oysters, combined perhaps with obscure environmental changes, so that the animals became locally extinct. Similarly the stocks of the free-living scallops and quins, which are caught by trawling, are threatened by over-fishing to supply the market for canned or frozen luxury sea-foods.

The capture of the larger crustacea for human food has already been noted in Chapter 3 where the danger to the stocks through over-fishing was pointed out. Nevertheless the smaller sorts such as shrimps and prawns, which are taken in huge quantities by the trawl, seem still to be in good supply, though industries that use them in bulk for manufacturing prepared foods and feeding-stuffs for domestic animals may be a menace to their continued abundance. A minor exploitation of the crustacea is shown by the use of barnacles as food, generally by the luxury trade. The meat of large acorn barnacles is relished in those parts of the world where they occur of sufficient size, and the stalks of ship-barnacles are a delicacy for gourmets in southern Europe.

Apart from the sea-cucumber or trepang already described in Chapter 3, the echinoderms, the sea-urchins and starfish, are of only minor importance as human food. Some of the sea-urchins are locally valued because their gonads make an agreeable bonne-bouche to certain people. In Great Britain the large sea-urchin, the "whore's eggs" of the fishermen, though named scientifically *Echinus esculentus,* are not generally eaten, though they suffer much depredation for their shells to be made into souvenirs for tourists.

Few invertebrates besides molluscs and crustacea are used as food by man, and where they are the exploitation is local and appears to inflict no damage on the stocks. A bizarre example is the use of the palolo worm by the natives of Samoa and Fiji. The palolo is similar to the rag worm of European coasts, much used as bait by fishermen, and lives in the crevices of coral reefs about two fathoms down. On the day of the last quarter of the moon, and the day before, in October and November every year, the worms leave their crevices and swarm to the surface in vast quantities to spawn. Each swarming-day makes a festival, whenthe natives also swarm out to gather the harvest of worms which are either eaten alive or baked, and are esteemed as a great delicacy. Worms with similar habits are also found on the coasts of other Pacific islands and are likewise exploited. The chief importance of marine worms to man, apart from their use as bait, is, however, indirect, because they form an important part of the food of many species of commercially valuable fishes.

The products other than food obtained from the invertebrates are not many, and few of them supply necessities, for most of them are merely articles of luxury or display. The fisheries for pearls and pearl shell were discussed in Chapter 3 – synthetic materials have largely taken the place of natural shell for making buttons. Similarly synthetic foam plastics have usurped the place formerly held by sponges in man's personal ablutions and those of his motor cars. The precious coral of the Mediterranean, which is not closely related to the reef-building corals, was formerly over-fished, but is being ousted by the innumerable synthetic substitutes for semi-precious stones. So, too, synthetic dyes have taken the place of those once obtained from certain gastropods, such as the famed Tyrian purple of the ancients; other man-made substances have replaced the Indian or Chinese ink prepared from the contents of the ink sac of cephalopods. A strange use of invertebrates lingered on into the medical world of the early decades of this century: the use of living leeches applied to bruises and black eyes, the last fading memory of the phlebotomy and blood-letting that formed the basis of much medical treatment up to a century before.

Although man's exploitation of the invertebrates other than insects is of no great moment apart from those he uses as food, the same cannot be said of the exploitation of man and his works by the invertebrates. Man is constantly surrounded by armies of hostile invertebrates bringing disease and death to him, his domestic animals and plants, and destruction to his artifacts. At the lowest end of the scale the protozoa, the minute unicellular animals that abound almost everywhere, include parasitic species that during the ages have killed millions of people. The blood parasites causing malaria, sleeping sickness, typhus, and other diseases have already been mentioned in the last chapter in connection with the insect vectors that carry them from one victim to another. Others, such as that causing syphilis, though not transmitted by insects, are equally destructive; as also is the parasite that lives in the intestine and causes amoebic dysentery. Many species, especially the coccidia and gregarines of the sporozoan protozoa are parasites of other invertebrates – the former live inside the cells and the latter in the gut of numerous species. The occurrence of these parasites in animals inimical to man offers no consolation for the damage he suffers, because most of them seem to have come to terms with the hosts from which they take their food, so that although they tax them they do not tax them to death. In spite of all man's ingenuity in making new chemical compounds for the treatment and prophylaxis of the evils brought upon him by these organisms there is little prospect that the survival of this part of wildlife will ever be endangered.

The parasitic worms, infinitely more complex in their structure than the protozoa, are animals that man might think he oculd well be rid of, but without them some as yet unrecognised ecological balance could be upset to his disadvantage. They belong to two zoologically unrelated groups, the flat and the roundworms, the platyhelminthes and the nematoda. The tape-worm and the liver fluke are familiar examples of the flat worms, at least in books, for what percentage of an urban population has ever seen living specimens of either? Both tapeworms and flukes usually pass through a complicated life-cycle, and in their early stages live in intermediate hosts before becoming adult in their definitive ones.

The liver fluke, which lives in the bile duct of sheep and other animals, including man himself occasionally, sheds eggs that pass out of the intestine in the faeces. The larvae that emerge from those that happen to be deposited in or near water swim about until they find a freshwater snail, into the tissues of which they burrow. Here they live for some time and produce further larvae of different form, which leave the snail and fasten to the stems of damp grass where they become encysted. If they are then swallowed by a sheep they continue their growth and migrate into the bile duct where they become adult. Flukey liver is commonly found in sheep and cattle when they are slaughtered; mild infection produces little obvious harm to the host, but a heavy infection produces poor condition, wasting, and death.

A large number of flukes are parasites of vertebrates, from fishes to mammals, and one in particular is dangerous to man in Africa. The adult flukes, named *Schistosoma* or *Bilharzia,* live in the big vein of the liver and the blood vessels of the bladder, causing pain, stone, haematuria, debility, and frequently death. The eggs pass out with the urine and the larvae that hatch bore into the tissues of a freshwater snail which they leave on reaching a further stage of growth, whereupon they swim freely in the water. These minute larvae bore through the skin of anyone bathing or wading, and are carried by the blood to their final destinations. The infection brings much suffering and debility; apart from drugs for the direct treatment of the symptoms, molluscicides to kill the snails and thus break a link in the chain of hosts are used in combating the parasites. Molluscicides, however, are unselective in their action and may produce unwated side-effects, by killing species other than the guilty vector.

The flat worms, or cestodes, also generally have life cycles that need an intermediate host. The adult worm, often of great length, lives coiled up in the intestine of its vertebrate host, attached by suckers or hooks, or both, on the minute head. The body consists of numerous joints each containing hermaphrodite reproductive organs, so that when ripe they are filled with fertilised eggs. The ripe joints break away and are passed with the faeces, from which the eggs may be dispersed to contaminate herbage with which they are swallowed by the second host. The eggs hatch in the second host and the larvae bore through the intestine to the muscles or other organs, where each forms a fluid-filled bladder, with a head like that of the adult tucked in from the surface. If the flesh of the host containing bladder-worms is eaten raw by the other host the bladder is discarded but the head which emerges from its pocket attaches to the lining of the intestine where it grows the joints forming the long body. The pairs of hosts, intermediate and final, range widely through the animal kingdom; pig and man, cattle and man, fleas and dogs and cats, rabbit and dog, sheep and dog, mites and rabbit, mites and cattle, fish and man, birds and water fleas, domestic fowl and house fly, and many others. Normally neither stage of the parasite appears to have any serious effect upon the host if the infection is not heavy, though feeding a large number of intestinal worms strains the host's metabolism.

It is the bladder-worm stage, however, that may sometimes cause serious damage, especially those species of bladder-worm that grow not one but many heads. If the bladder-worm derived from a dog tapeworm lodges in the brain of a sheep, the intermediate host, it grows to such a size that it compresses the brain, and the animal's muscular co-ordination is consequently damaged. The "giddy sheep" suffers from the staggers, cannot feed, and dies unless swiftly hustled to the butcher. Man sometimes gets infected with bladder-worms by eating contaminated food or by self-infection, thus acting as intermediate as well as final host. If he is so unfortunate the bladders may lodge in the brain, lungs, liver or other vital organs and grow large enough to form hydatid cysts causing serious illness or even death. It is surprising that accidental infections from dogs, cats and other house pets whose fur may be contaminated with tape-worm eggs are not more common in those, particularly children, who fondle them.

Tapeworms are a form of wildlife that man is unlikely to exterminate or even endanger, much as he might wish to. Although the odds against any one egg becoming an adult are astronomical, the eggs are produced on a vast scale and are widely dispersed because of their minute size. It is difficult to see how any unwanted side-effects damaging to the environment could be brought about were it possible to exterminate tapeworms; in both stages of the life cycle their relationship to their hosts is entirely in their interest, nor are they the food of any other species. It is unlikely that they have any appreciable effect in keeping population numbers of their hosts in check, except possibly in rare instances as in adding to the mortality in some abnormal and overcrowded situations.

There is an exception to the uselessness of cestodes as food for other animals. The larva of a tapeworm found in some water birds lives in the body cavity of the stickleback, not as a bladder-worm, but as a ribbon-like many jointed worm. In some parts of Europe the worms removed from the stickle-backs are or were regarded as delicacies for gourmets, to whom they were served up as "vers blancs". In 1741, during his journey to Gotland, Linnaeus found an infected stickleback; when he squeezed it "out came a big white worm" and he was surprised that there was room for it in so small a belly. He did not eat it.

The nematodes, or round worms, are cylindrical and unjointed; they range in size from the parasitic kidney-worms of whales, several feet long, to minute thread worms that can be seen only with a microscope; they are parasitic in animals or plants, or live freely in soil or water. The parasitic kinds trouble man because they are damaging to his health and that of his domestic animals, and destructive to his cultivated crops. In vertebrates the worms generally live in the intestine where they cause irritation and inflammation, excrete harmful toxins, and where some species suck the host's blood. The minute eggs of the worms pass out with the faeces and contaminate the ground, herbage, bedding, skin and fur, or clothes of the host so that the sufferer is continually reinfected. It is almost impossible to keep domestic animals entirely free from parasitic nematodes, but veterinarians and farmers dose stock with vermifuge medicines to keep the worm-load of the animals down to a tolerable level, which is determined by counting the eggs in a unit sample of droppings. Until the invention of modern chemical vermicides the losses of stock from infection of the intestine, lungs, blood vessels, heart and kidneys by different kinds of nematodes were enormous; poultry, too, were killed in thousands from infection of the windpipe by the fork worm – a nematode which causes the fatal "gapes".

Although the common round worms and thread worms living in the intestine of man, and especially in that of his children, cause discomfort, it is three other kinds that are dangerous to his health and life. The hook-worm, common throughout the warmer parts of the world, lives in huge numbers in the intestine, where it damages the blood vessels, causing anaemia and death in the pot-bellied victim. The minute blood parasite, filaria, which causes elephantiasis and other symptoms, has already been mentioned in Chapter 7, because it is transferred from host to host by a mosquito. The third dangerous nematode, so small that it is barely visible, is *Trichina,* which lives in the intestine and releases enormous numbers of eggs. The larvae which hatch from them bore through the wall of the intestine and are carried by the blood to all parts of the body until they come to rest in the muscles, where they form tiny cysts in which they remain coiled up and may live for years. If trichinised muscle is eaten by another animal the cysts release the larvae in the intestine, where they become adult and lay eggs to repeat the cycle.

The usual host from which man is infected is the pig, which is itself infected by eating raw pork offal or the bodies of rats, a species also susceptible to infection; it must be rare indeed for trichina to be passed from man to animal. During the Second World War a widespread outbreak of infection in Great Britain was caused by the war-workers' habit of using the contents of uncooked sausages as a sandwich spread – sausages made from infected pork imported from South America. Thorough cooking kills the parasites. The larvae cause degeneration of the muscle fibres in which they were encysted, but the worst effect is the fever which occurs when the larvae enter the blood stream and are carried throughout the body. The guinea worm, which lives in cysts under the skin, usually of the legs, of people in hot countries, causes painful tumours and may be fatal if its body wall is accidentally ruptured so that the contained embryos pass into the blood of the victim. Normally the embryos escape into water where their intermediate host is a water flea.

The nematodes parasitic on plants are equally destructive to man's interests; many are minute but swarm in the soil, feeding in roots or other tissues and thus causing reduced yields or complete destruction of crops; some species even live in and destroy the ears of cereals.

The nematodes form a part of wildlife that is largely unseen by urbanised man, who indeed is hardly aware of its existence. Yet the worms are everywhere, often in great numbers – no vertebrate animal, and probably no plant is free from them, and although infection with some species appears to be of little importance the ravages of others can be disastrous. The free-living kinds, generally minute, especially those that live in the soil may, on the other hand, be indirectly useful to man by breaking down organic matter in the formation of humus.

Of the other invertebrates that attack or parasitise man we may notice the leeches, especially the land leeches of damp tropical forests which, although only about an inch long, occur in such ferocious swarms that they produce a more than trifling loss of blood, much irritation, annoyance and disgust. Their bite is so gentle that it is often unnoticed, but the wound left may not only become inflamed but infected and ulcerous. The same result may follow the bite of the common harvest bug, the young stage of one of the mites – the lesions are sometimes called 'heat-bumps'. In the tropics the bites of the young stages of various kinds of ticks are intensely irritating to the victim; in addition, as already noted, some ticks are the vectors of diseases such as relapsing fever in man and red-water fever in his cattle.

Two minute species of mite live parasitically in the skin of man and other animals: one lives in the hair follicles where it produces follicular mange, the other, the itch-mite, burrows under the surface of the skin producing scabies or sarcoptic mange. Mites are also destructive to crops: every gardener knows the havoc that can be wrought by the mite that he calls red spider, though many other less conspicuous species are equally as damaging. Another mite, rather similar to the follicle mite, lives parasitically in the buds of the blackcurrant where it produces 'big bud' with a resulting loss of the crop.

The tick vectors of disease actively introduce the infection into the animal they bite, but other invertebrates can act as passive vectors, especially some of the bivalve molluscs eaten by man. These shellfish are filter feeders that extract minute organisms down to the size of bacteria from the water. From time to time some of the blue-green unicellular planktonic algae reproduce in enormous numbers, making the water thick and soupy and producing a 'bloom', formerly known to the North Sea trawlermen as 'Dutchman's baccy juice'. The algae produce a toxin which is dangerous to many animals including man, so that if he eats shellfish contaminated by a bloom he is poisoned. Within the last few years a bloom in the North Sea off the English east coast killed some people and great numbers of sea birds and fishes in this way. Such shellfish can also be contaminated with the bacteria of disease where untreated sewage is discharged into the sea – gastro-enteritis, typhoid and cholera may thus be distributed. If man catches these infections in this way one can only say it serves him right, first for his filthy habits of sewage disposal, and secondly for his foolishness in eating shellfish from places known to be contaminated.

The nematode worms and the mites have already been mentioned as pests of crops, but other invertebrates join in the onslaught upon the abundant food provided by man in his gardens and fields, though they are perhaps of minor importance. The damage inflicted by land snails may be distressing, but is far exceeded by that done by the small slugs that live in the soil and emerge by night to feed, and are particularly destructive to seedlings. A greater menace comes from the arthropod millipedes known as wire-worms, which feed on plant roots; the centipedes on the other hand are carnivorous and may be regarded as beneficial by feeding on insect pests. Among the crustacea, which are typically aquatic animals, the wood-lice are an exception in their adaptation to terrestrial life and to breathing air; their destructive appetite for the gardener's plants is to some extent offset by their recycling of decaying vegetable matter.

The robber crab of tropical coral islands is another crustacean adapted to terrestrial life. It is more closely related to the hermit crab than to the ordinary crabs and lobsters but is large as crabs go, being a foot or more long, and is one of the most bizarre crop pests. It feeds on coconuts, not only those that fall to the ground , but it actually climbs up the palms to pick the nuts from the crown. Mention has already been made of the invertebrate pests of the oyster farm – carnivorous molluscs of several kinds, together with starfish; and of the destruction of coral reefs by the exploding population of the crown of thorns starfish.

Numerous other marine invertebrates set man worse problems to solve – the ship-fouling organisms, and those that destroy his artifacts. In addition to many kinds of invertebrates, the sea-squirts or ascidians, which are lowly members of the chordate phylum, together with sea weeds, grow rapidly on the underwater part of a ship's hull, producing a thick blanket that offers great resistance to passage through water. In the days of sailing ships the fouling greatly reduced the vessel's speed; it was removed by careening the ship – putting her on the hard in tidal waters and heaving her down on one side and then the other so that the growth could be scraped and burnt off and the planking re-payed with tar. In later days, when sailing ships were larger, cleaning the hull was done in dry dock. When steamships came in, bottom-fouling not only reduced speed but also increased fuel consumption, a matter of even greater importance to the owners. All the fouling organisms have planktonic young stages which swarm in the water ready to settle on any suitable substrate; although barnacles, both acorn and stalked, form a large part of the fouling, a well-fouled bottom is a rich biological museum – or zoo – of many species. Numerous anti-fouling paints have been invented, and much research and money have been expended in producing them, but no complete and permanent preventer of fouling has been found. Anti-fouling paints contain poisons to kill the foulers as they settle; the poison must slowly leach out into the water, and when it is all expended the paint must be renewed.

Fouling problems are not confined to ships; power stations drawing cooling water from the sea are plagued by fouling organisms choking their culverts. The common mussel is often the main culprit, for the constant stream of water passing over them brings a continuously renewed supply of food so that they flourish exceedingly, but many other animals including sponges help to make the obstruction. Water-supply undertakings far from the sea are also sometimes similarly embarrassed by the growth of freshwater mussels and freshwater sponges blocking their pipes.

In addition to the foulers there are the borers, animals that feed upon and destroy artifacts made of wood immersed in water. The most important are a mollusc, the shipworm, and a small isopod crustacean, the gribble. The shipworm is a bivalve with a long naked worm-like body and very small shells at one end, which are used as jaws to rasp away the wood. The larva settling on the surface of wood bores in, leaving a minute pin hole; as it grows in length it eats away a tubular borrow lined with a shell-like substance, and can reach a length of a foot in less than a year. The wood may be so infested that it is almost eaten away and replaced by a mass of fragile tubes, though appearing sound on the surface because the holes of entry are so small. Many a wooden ship has been destroyed by shipworm; Drake's "Golden Hind", after circumnavigating the world, sank at her moorings off Greenwich because her planking was riddled by shipworms. Wooden ships were coppered to prevent the worm attacking, as they are still today in some parts of the world, though alloys such as monel metal are now more often used. A ship was copper-bottomed by nailing thin plates of copper all over the underwater part of the hull, usually with a layer of canvas and tar sandwiched between.

Shipworms not only attack ships, driftwood and other floating timber, but piles and the woodwork of piers and sea defences in coastal waters, where the salinity is not too greatly reduced by the run-off of freshwater from the land. These structures are also attacked by countless numbers of a small crustacean, the gribble, more destructive to immersed timber than the death-watch beetle is to the roof timbers of an ancient church. Man has gone to immense expense and trouble to combat the pests by preventing their access to his structures. Various poisonous compounds are now used, but the former painfully laborious method was to scuppernail the timber. Scup-head nails have very large heads and they are driven into the timber close together so that the heads overlap to form an almost continuous protective covering, in which the interstices are filled by the rust which quickly forms and welds all together. Stone constructions, too, are attacked by boring animals, such as the piddock and other bivalve molluscs, some sea urchins, and even certain sponges; but the damage done to man's interests by the stone-borers is insignificant compared with that inflicted by the destroyers of timber.

The insects, as described in Chapter 7, are the main invertebrate group with members that destroy or damage man's artifacts on land, but the mites also inflict serious losses. The mites are arachnids, related to the spiders, small in size and living on land or in either fresh or salt water. Some microscopic species of mite that feed on decaying or dried organic matter are often a nuisance, and sometimes a serious pest to man. They make up for their small individual size by their enormous numbers, as can be seen if a cheese infested with cheese mites is examined with a hand lens. In addition to those species that destroy or spoil stored foodstuffs there are others that infest houses, where they feed on all kinds of dry debris, and sometimes infest upholstered furniture in such swarms as to leave it fit only for the bonfire.

Finally there is a host of invertebrates that have unpleasant ways of defending themselves if they are inadvertently molested. They do not seek out man or his works to attack them, but are of a kind with the *'bête bien méchant, quand on l'attaque il se défend.'* Sea bathers get stung by jellyfish or weever fish, and on rocky shores in the tropics may get their limbs filled like pin cushions with the long spines, sharp as needles, of certain kinds of sea urchin. On tropical shores, too, the collector of sea-shells must beware if he seizes one of the beautiful and highly prized *Conus* species. In these gastropods the radula, the ribbon-like rasping tongue, carries a number of comparatively large, pointed and barbed teeth connected with a poison sac – if disturbed the animal bites in a flash, inflicting a painful wound which is said to be sometimes fatal.

On land the bites of large centipedes, and the stings of scorpions, can have serious consequences for anyone unlucky enough to receive them; travellers camping in the tropics are careful to shake out their boots every morning before putting them on. The bites of various species of spider are reputed in several parts of the world to be dangerous to man, because the venom injected by their fangs is unusually toxic. The tarantula of southern Europe, whose name is wrongly transferred to the tropical 'bird-eating' spiders, was said to inflict a bite causing madness and hysteria, curable only by wild dancing until the victim was exhausted. The identity of the species is uncertain, and the production of the symptoms by its bite is doubted; the tarantella, however, remains.

There is no uncertainty about the black widow spider of America, which inflicts a dangerous and painful bite – usually on the user of a country earth-closet, where the spider lurks under the seat and makes its home. As modern sanitation progressively takes the place of the works of Mr Lem Putt, black widow bites on innocent bottoms may become rarer. Examples of so-called noxious invertebrates such as these can be multiplied many times, but man cannot rightly regard them as enemies because although they are armed they remain neutral unless accidentally disturbed.

One may well ask whether there is anything to be said in praise of invertebrates and their relationships with man to counterbalance the long catalogue of parasites, spreaders of disease, and despoilers of crops and man's artifacts – stingers, biters and blood-suckers. Apart from those invertebrates that provide man with valuable and often delicious food, the great web of invertebrate life is important to man because it is the basis of the environment's metabolism. In the oceans the invertebrates are the main links in the food chain leading from the primary producers, the planktonic photosynthesising diatoms and other microscopic plants which use the energy of sunlight to build organic from inorganic matter, to the vertebrates, from the smallest fish to the great whales, the sea birds and most of the marine reptiles. On land the teeming microscopic invertebrates of the soil are ever at work recycling organic debris and, together with the bacteria, producing the humus in which plant life can flourish and fulfil its role as the primary producer by harnessing solar energy. Among the larger invertebrates of the soil the earthworms, as Darwin showed, are the natural ploughmen, bringing up huge quantities of humus which becomes dispersed over the surface, and aerating and draining the topsoil at the same time. The lugworm living in the sand and mud of tidal waters performs a similar service in the sea.

The invertebrates, although small and inconspicuous, have also played a part in the formation of the physical as well as the biological environment, a part that has not been remotely approached in results by man, for all his colossal earthmoving equipment. No doubt his embankments and canals, motor roads and railways, dams, quarries and mines will still scar the earth for many millennia, but these are mere details in a landscape much of which has been formed by invertebrates.

They have built the coral reefs of the tropics and the atolls strewn over the vast area of the Pacific ocean, which served as stepping stones for man in his dispersal through the southern hemisphere. The corals, though classified on a low level of the taxonomic scale, lie far higher than the protozoa, some of which have formed a part of the very rocks of the earth's crust. In Great Britain the chalk formation, from the white cliffs of Dover through the southern downlands, the Chilterns, the foundations of East Anglia underlying the boulder clay left by the Great Eastern glacier, to the worlds of Lincolnshire and the East Riding, is a former marine sediment made from the shells of unicellular planktonic protozoa living in the clear waters of a shallow sea, though some of the chalk may be the result of chemical precipitation. The flint scattered through the chalk was formed from silica derived from the spicules of sponges and the remains of other siliceous organisms. At the present day a similar sediment is accumulating at great depths on the floor of the Atlantic ocean by the deposit of the limey shells of the protozoan *Globigerina* to form the ooze named after it.

The influence for good or ill of the invertebrate part of wildlife has always been far greater on man than has man's upon the invertebrates; in spite of the efforts to control them in his attempts to manage his environment it is likely that for the greater part they will remain unconquered.

9 PROFIT AND LOSS: CONSERVATION

In the preceding chapters we have examined some of the ways in which man and wildlife, both planrs and animals, affect each other. Although man's alteration of the environment has profoundly influenced wildlife, he has himself been subjected to equally far-reaching influences from wildlife, so that he cannot isolate himself from the ecology of the world. We may now briefly recapitulate some of the more important points ot our survey, and go on to consider the probabilities of man being able to find a way of living peacefully with wildlife; for it is certain he cannot live without it.

Ever since the first appearance of life on earth a process of evolution from comparatively simple to more complex organisms has been going on. Evolution, however, did not begin with life, for living matter did not appear suddenly but itself slowly evolved from the non-living. It would indeed be difficult to draw the line between living and non-living matter during the process – is a crystallisable virus living in its crystalline state, or a molecule of deoxyribonucleic acid, the basis of the genetic code, alive or non-living?

In the complicated three-dimensional web of plant and animal life each species is in contact with many others, influencing them and subject to their influence, to make up the ecological pattern found in every environment. Nothing, however, is static; apart from the effects of different organisms on each other, wildlife has throughout geological time been subject to climatic changes, continental drift, the rising and sinking of land masses, eustatic changes of sea level, and the alterations produced on the surface of the earth by the processes of plate tectonics.

No two living organisms of the same species are exactly alike, except identical twins or higher numbers of siblings among animals, and clones among plants, which are derived from a single egg cells. Individual variation, obvious to man among his fellows, but not so readily apparent to him among other species, can sometimes by chance produce forms better able to survive in changed conditions. If the variation is hereditable it may give rise to a new population, providing the genes producing it are not lost among the gene-pool of the species. They will be so lost unless the organisms carrying them breed only among themselves and are isolated reproductively by geographical barriers, behavioural or biochemical differences, and so on. Such isolation can equally perpetuate fortuitous differences that neither favour nor threaten survival. Since there would be no elimination of individuals carrying new characters that brought no handicap, much of the infinite variety we see through out wildlife may be quite meaningless in the context of evolutionary adaptation. Some characters of every species fit it for its way of life – unless it adopts a way of life to fit its characters – but many of them may be entirely indifferent, and without survival value. If, however, a variation is favourable to survival the population carrying it may replace that lacking it by differential mortality or reproductive success.

The process unfortunately termed 'natural selection' – for nothing is positively selected, it is merely the less fit that perish – together with fortuitous changes in the genetic code, produced the almost infinite variety of wildlife that we see around us. The correlation between form, function, and the exigencies of the environment is generally so close that the organism's adaptation seems to have been moulded by the environment. In the early nineteenth century this coincidence gave rise to the erroneous belief that characters acquired by organisms during their lifetime in adaptation to the environment were inherited; but new characters cannot be inherited unless they are the expression of changes in the genetic code. The code, which is written in the sequences of nucleotide bases in the molecules of deoxyribonucleic acid (DNA) carried by the chromosomes of the cells, ensures that the characters of species remain stable unless the code is changed.

When new forms replaced older ones it is suggested that they were better adapted to the environment, but adaptations even in the old species appear to be so perfect that it must be assumed the new forms were adapted to a changed environment, and in this sense it can be said that the environment moulds the species. The species that once flourished, but became extinct and were superseded by new ones, far outnumber those now living – the extinction of wildlife on an enormous scale is necessarily part of the process of evolution.

There is no reason to think that evolution has stopped because we see little change in the character of the world's biomass by the appearance of new species by hybridisation and polyploidy, the multiplication of the numbers of characteristic chromosomes. Many of the cereal cultivars are such species – the cord grass *Spartina townsendi* is another well-known example. The evolution of new species of animals, before our eyes, through changes in the genetic code is not apparent because it is believed to take place by natural selection acting upon cumulative small changes over a long period of time, in populations isolated by geographical or other barriers. Evolution thus always appears to have taken place in the past but to be no longer operating in the present, though that past is not always remote. The volcanic island of Hawaii carries a fauna of numerous related species of land snails, isolated from each other in its precipitous valleys, and also has many species of fruit fly isolated in its forests; furthermore the island has or had a number of endemic species of birds, notably the ne-ne or Hawaiian goose. As the Hawaiian islands are geologically of recent origin, all these animals, together with the endemic plants, must have evolved in a comparatively short time from the ancestral species accidentally brought by wind or wave. The genetic changes in these isolated populations have accumulated until the populations have become separate species.

A similar but less clear-cut evolution is shown by many species that have a wide distribution, so that breeding generally occurs between near neighbours of the population in the different parts of the range; this isolation results in the production of subspecies and clines. The subspecies at opposite ends of a cline may differ markedly, but they are connected by intermediate forms that differ only slightly from their immediate neighbours. Subspecies can, but do not usually, interbreed except where their ranges meet or overlap, as for example do the carrion and hooded crows. The different breeds of domestic animals, too, are not distinct species; they are produced by artificial selection ringing the changes on the already existing genetic material, with mutants appearing when a change occurs in part of it.

Man, like the rest of wildlife, has similarly evolved, from the first appearance of primitive hominids two million or more years ago to the emergence of modern man some 30,000 years ago, when he replaced Neanderthal man in Europe. Neanderthal man was black, and when he moved into temperate regions following the retreating glaciers he died out from vitamin D deficiency, as shown by his crooked bones, because ultra violet light, weaker than in the tropics, could not penetrate his pigmented skin to help make the necessary vitamin. Cro-Magnon man, our ancestor with less pigment, took his place. As we have seen in earlier chapters he was for long merely part of wildlife's ecological web before he began to make an increasing impact on the environment and its wild life. Like all wildlife he exploited the environment for food, and like some wildlife he also exploited it for shelter and territory, the more successfully when he took to living in social groups larger than the family party. His society, unlike that of the social insects where the individuals are entirely subordinate to the needs of the group, preserved individuality while cooperating with the rest, a type of relationship encouraged by communal hunting and shown by other carnivorous mammals – the sight of a pack of African hunting dogs chasing a gazelle brings to mind nothing so much as a football team punting a ball about.

The invention of language, permitting much more precise communication than that enjoyed by other creatures, and of private property in the form of clothing, tools and so on, however crude, emphasized individuality, and led to further exploitation of wildlife and the environment to satisfy new needs. Man drew further away from wildlife as his civilisation developed and he increasingly made an artificial environment for himself. He likes to think that his artificial environment gives him freedom but he deceives himself, for he has made it so complex that he is a slave to it and as helplessly dependent upon it as an infant is upon its mother. If there is no petrol in the pumps, no milk and bread on the doorstep, no meat or groceries in the supermarket, civilised man will starve because he has forgotten how to fend for himself; the social order will break down, rioting will break out and people will start killing each other – except the minority of the ruling classes who have, at public expense, made preparations for what they call an emergency, and will retire to the deep and well-provisioned bomb and radiation-proof underground towns that are already built and waiting for them.

The fragile environment which man has spun round humself like a cocoon aims at freeing him more and more from the need to work. Machinery and automation, large-scale engineering, and factory farming increasingly serve all his needs and give the masses leisure time they have never had before. Leisure, however, in more than small quantities, is quite unbiological. The biological purpose of wildlife and of man, if they can be said to have any purpose, is to find enough food and to reproduce – among the higher vertebrates, only when the daily ration of food has been found and eaten can leisure be enjoyed in indolence, as in the old maxim for good health, 'Live on sixpence a day, and earn it.' Work is a biological necessity, and when man is no longer forced to work he has too much leisure, gets bored, and has to invent pseudo-work to exercise his mind and muscles, or at least his muscles.

From the earliest civilisations to the present time people with leisure have filled it with sport, games, war, the arts and other activities. Large parts of the population are now obsessed with games, either as spectators or participants, with gambling, field sports and other unproductive pursuits to stave off the boredom of leisure. This enormous expenditure of energy and material resources is biologically futile, for once an individual has produced enough offspring to replace their parents his biological purpose is fulfilled; he is of no further use and competes for resources with his younger fellows who are still biologically "significant". Yet, as Ambrose Smith said long ago on Mousehold Heath, life is sweet, so that man goes on trying to gain material and mental pleasure until his last gasp.

In satisfying his needs – whether real or artificial – for ever more possessions and amenities man has from the earliest times been inventing new and better ways of exploiting wildlife and the environment to obtain them. For centuries this was thought to be entirely laudable, and when it brought a reduction in returns, or the near extermination of a wildlife species, the loss was generally regarded with indifference and accepted as what is called in modern jargon 'a fact of life' or 'one of those things'. The industrial evolution of the nineteenth century, which quickened its pace during the twentieth through the rocketing achievements of science and technology, increased the pressure on the environment so enormously and brought such vast changes and devastation, that almost suddenly civilised man has seen that what he is doing is no longer praiseworthy pioneering, but threatens to destroy himself as well as the rest of wildlife.

The environment and the lives of some of its more conspicuous wildlife inhabitants are being changed rapidly in ways detrimental to them and to man himself; man is using up resources that are not renewable, and exploiting those that are, far too fast. The exploitation, together with pollution, especially of the seas and fresh waters, both with substances intended to control part of the environment but used recklessly and in excess of needs, and with garbage and industrial wastes, results in the destruction of much wildlife and threatens some of it with extermination. When the world human population was smaller than it is now the use of the sea as a universal *cloaca maxima* did little harm because the wildlife, from bacteria upwards, was able to break down and recycle the organic matter which was not contaminated with powerful poisons. Today the dumping and discharge of unwanted matter is excessive; we know how to make most non-polluting, but are unwilling to pay for the necessary treatment and think to solve the problem by putting it out of sight. Meanwhile the wasting assets of the world continue to waste.

Now that man at last begins to see what is happening he is frightened; he sees that many things he thought would be there for the taking in inexhaustible supplies are rapidly running out, and he fears for the future. The thought that much of wildlife, too, will soon disappear smites the conscience of those who find pleasure in the contemplation of wild nature, so that they are moved to attempts at preserving it. Just as man has in the past unthinkingly destroyed, some who are swept up in the propaganda for conservation unthinkingly wish to put the clock back, although they well know that time goes in only one direction. The more hard-headed supporters of conservation know that we must accept changes, but hope that we may be able to guide them in directions that will not completely sterilise the environment. The human population explosion, however, always threatens to stultify their efforts: there are too many people, they put too heavy a demand on resources, and compete with wildlife for living room. The problem may be self-resolving because too large a population not only strains natural assets, but also social relationships; it engenders aggression so that a quarrel could at any moment, with the aid of modern agents of destruction, drastically reduce the size of a population at a stroke.

We must not forget, on the other hand, that though we may dislike change we accept the results of the changes of, say the last two thousand years with pleasure and no regrets. Great Britain was covered with forests two thousand years ago, and would be today had not man destroyed them, yet nobody deplores that they have gone. Similarly the clearing and improvement of land for agriculture, the building of cities and roads, the extermination of bears, wild boars and wolves, and in more recent centuries the draining of the fens, and the agricultural enclosures which give us our much admired countryside, are all accepted as normal and desirable – and most people do not know how artificial the English landscape is.

Just as man's physical characteristics, like those of all living things, are determined by his genetic constitution, so also are his mental characteristics, his ability to learn, to reason and to adjust or alter his behaviour in the light of circumstances: he cannot be better than his genes. His behaviour towards wild life, and his treatment of it, must therefore be determined by his genetic code; behaviour that was formerly almost universally destructive, but is now becoming conservative among some parts of his population. Man's destruction and pollution of the environment must thus be regarded as part of his natural evolution, as also must his growing disquiet about the results of his activities – expressed in the desire of some, at present a minority gaining increasing sympathy with its ideas, to call a halt to destructive exploitation by changing to conservation policies. The adoption of those policies is, however, hampered by the ever-growing pressure on the environment brought about by the population 'explosion', the great obstacle to improvement that constantly frustrates every proposal.

However good people's intentions may be, especially when they wish to stop somebody doing something they do not themselves want to do, the highest principles tend to be thrown out of the window the moment their personal interests are adversely affected. We see this plainly in the 1973 fuel 'crisis'. The threat of a dearth of petrol, and of oil for heating and for making electricity, and the prospect of rationing, brought immediate world-wide proposals for the relaxation of the regulations to reduce pollution both in the air and the sea, and the conservationists' opposition to the building of the Alaska pipeline is over-ruled with the tacit consent of the majority.

Only a year after the ban on the use of DDT in the United States the Department of Agriculture backed an application by the lumbermen to have the ban lifted so that DDT can be used to kill the caterpillars of the tussock moth, which are devastating the pine forests of the north-west. The lumber-men claim that the damage to the environment by the population explosion of caterpillars will be greater than that caused by the persistent pollution resulting from the use of DDT; the financial loss through spoiled timber will be great if the infestation is not stopped.

In Great Britain, and especially in Scotland, the legal protection of grey seals has had unexpected results. Protection of the seals from commercial exploitation turned out to be unnecessary because the small-scale hunting of the nineteenth century came to an end with growing human prosperity and the decline in peasant subsistence farming and fishing; the number of seals was too small to attract the attention of commercial interests. Thereafter the population increased exponentially and in recent years almost suddenly became many times greater than before. When the damage to fisheries by seals became conspicuous, because seals eat salmon and spread the cod-worm, commercial pressure was such that the protection was lifted to allow a kill of young seals in order to reduce the population – a kill euphemistically called a "cull" to pacify sentimental public opposition. In disposing of the unwanted carcases of the slaughtered seals officialdom unintentionally encouraged a market for sealskins; prices shot up, and licensed killers now reap a harvest. Why should they not, if their productivity fills a market demand without endangering the stock that provides the crop?

The number of alligators in the United States was enormously reduced by over-killing so that the animals were given complete protection, and the use of alligator skins for making leather goods was made illegal. In a few years the numbers had greatly increased and the animals, unharried by hunters, grew bolder than before and took to eating fur-farmers' stock, domestic pets and even killed one human bather. Some of the States have therefore lifted the protection, and propose to allow a kill of some thousands of alligators in the hope of mitigating the nuisance.

The protection of apparently harmless creatures can also 'backfire' in unexpected ways. Bird strikes on aircraft taking off or landing cause much damage and sometimes danger. In Great Britain, as noted in Chapter 5, the peewit is one of the species which, with gulls and others, find the vast areas of short grass on airfields convenient roosting or feeding grounds. The peewit has greatly increased in numbers since complete protection made the taking and marketing of plover's eggs for the delicatessen market illegal. The species may, indeed, be over-protected, for in the Netherlands where also it is protected, there is a limited open season for gathering its eggs – and much competition to find the first clutch for formal presentation to the Queen. The product of the second layings after the season has closed seems to be ample to preserve the stocks.

On the other hand the irrational harvesting of whales on a scale that has ruined a once profitable industry, and lost us a valuable source of raw materials, is rightly deplored by economists, though more loudly condemned by sentimental do-gooders who have never seen a whale, know little or nothing of the problems involved, and are swayed only by their emotions. Were they in the position of the Japanese, whose overcrowded mountainous islands have little land suitable for raising animal protein, and who consequently have relied for many centuries on the products of the sea, including large amounts of whale meat, for their diet they would be less vociferous in their demands for the total abolition of whaling. In Europe we use whale meat for making tinned food for pet dogs and cats, but the animal lovers seem able to balance their love for one species against that for others – kindness to animals means kindness to nice animals.

In many western countries brutal sports such as bull- or bear-baiting, cock-fighting, arbeits – unarmed combat between man and dog – badger-drawing and others are now illegal. Nevertheless badger-digging, cock-fighting and rat-killing are still enjoyed in Great Britain; and the traditional Spanish bull-fight is said now to be staged as a tourist attraction as much as an entertainment for the native afficionados. The popularity of violence on television shows that laws do not alter innate appetites, which may thus be vicariously satisfied. Those who might once have enjoyed the now illegal sports can have the thrill of watching television natural history films such as a pack of African hunting-dogs pulling down a wretched antelope and tearing out is entrails to eat it alive.

The expanding population, together with artificially stimulated demand give rise to the political fetish of the expanding economy, a looking-glass rat-race in which an ever-increasing speed is needed to stay in the same place.

Economies cannot expand indefinitely – there are limits to growth – and if only we could attain a stable population we could have a stable economy, without the frenzied urge to expansion that leaves people no time to relax and think of where they are going, or even to think where they really want to go.

The world, however, is not yet entirely a desert, nor is it so completely despoiled that its beauty has departed for ever. It is true that the population increase together with urbanisation and industrial progress has fouled the environment and brought about the local extermination of many kinds of plants and animals. Local extermination, however, is not necessarily total extermination, and for the greater part there is plenty of wildlife left in other places. Conservationists fear that many kinds of wildlife that have become scarce as a result of human predation, or encroachment on wild habitats through expanding development, will follow into extinction those species that are already lost.

It is pertinent to consider just how many species have become extinct in historical times. There are over 8,500 living species of birds, and during the last three hundred years seventy odd species have become extinct, some through natural causes but the majority directly or indirectly through the activities of man. Many of them were extremely vulnerable because they existed in only small populations, generally on islands, and were so closely adapted to their habitats that they could not survive the changes in it brought by man. The flightless rails, the dodo, and other ground birds could not survive when man's domestic animals were introduced into their habitats, and destruction of habitats by fire and felling left them with nowhere to live. Several species, however, are notable exceptions, such as the great auk, once plentiful in the North Atlantic, the passenger pigeon, and the Carolina parakeet, the last two existing in great numbers in North America into the nineteenth century, all reduced to the verge of extinction by man and now lost. On the other hand the extinction of species is part of the natural course of evolution; geologists have estimated that the number of bird species was reduced by as much as a quarter during the two million years of the pleistocene epoch, in which severe glaciations periodically extended from the poles towards the equator.

Among the mammals it is the larger kinds known as the 'big game' animals, and thus by definition tacitly assumed to be suitable creatures for man to kill for his amusement, that give the greatest concern to those anxious to save the species threatened with extermination. There are something over four thousand kinds of mammal living today, but how many of the big game species have been exterminated in the last three centuries? The answer is, remarkably few, although many have been greatly reduced in numbers. In Africa, the big-game hunters' paradise, two species have become extinct, the blue buck of the Cape about 1800, and the quagga, which had a much greater range, about 1880. Local extermination has lost us many subspecific races of species that live on elsewhere, though a few, such as the white tailed gnu, the blesbuck and the bontebuck, are now all enclosed by fences.

During the same period Asia has lost one species, Schomburgk's deer, which was last seen during the 1930s and is probably extinct, though it is possible that a few remain in remote parts of Thailand. North America has lost none of her big game, though the bison and the pronghorn had both gone far along the road to extinction before they were rescued by legislation. The only large Pacific Ocean mammal to become extinct in historic times is Steller's sea-cow.

Of the medium-sized mammals historically extinct the Falkland Islands fox or Antarctic wolf may be mentioned, as also the Japanese wolf, the sea-mink of North America, and the parma and toolach wallabies of Australia. For the rest the exterminations are those of smaller species including rodents, insectivores and bats, particularly of the West Indies and Malaysia, and some of the smaller marsupials of Australia where, until the last few decades, the human attitude to wildlife was either hostile, or indifferent to excessive commercial exploitation. Of the something over four thousand species of recent mammals the total lost in historic times amounts to about fifty, including some obscure rodents and insectivores which may have become extinct by natural means rather than through man's destruction.

We may thus look at the qualitative extermination of wildlife in perspective and see that all is by no means lost; a few mammals have gone, some birds, a few reptiles and some pretty insects, but compared with what is left it is only a tiny fraction, which for the most part consists of higher vertebrates. Only about 200 species altogether have become extinct in historic times, many by natural causes and not by man's influence direct or indirect, leaving a million named forms known to science still living with us, and many more yet to be recognised.

When we come to the quantitative extermination of wildlife the position for many species is much more serious, because a large number of species, particularly of the bigger mammals and birds, has been reduced to numbers so low that their continued existence is seriously threatened. If they could be left unmolested in what remains of their habitats they might have the chance of living on, and perhaps increasing to the limits of the support provided by their environment. This cannot be without deliberate action by man to help them. Not only must he refrain from killing them for food and other useful products, or for sport, but he must forgo taking their habitats from them for his own use. Protecting scarce species by passing laws forbidding people to kill them is easy; protecting habitats from destruction by an ever-increasing human population is not. The very term "developing country" implies that a growing population will bring the wilderness into subjection and replace wildlife with cultivated crops and flocks and herds of domestic animals, that roads and cities will be built and manufacturing industries will be established. If any larger kinds of wildlife are to remain, a deliberate decision must be made that large areas of the developing countries shall not be developed. Can the inhabitants be expected to agree?

The conservationists may quote in reply, 'Man shall not live by bread alone' – true, but with hungry mouths multiplying faster than supplies of bread, aesthetic arguments carry little weight, and the hungry are apt to think the aesthetic emotions of the well-fed are no more than humbug. We are brought back to the fact that there are too many people competing with wildlife for sustenance. Furthermore, the cause of conservation is not always helped by the fact that many of its patrons are notorious shooters, living in mansions festooned with big-game trophies; the cynical may say they are poachers turned game-keepers, and having had their fun they want to prevent others enjoying the same privileges. Nevertheless the patronage of 'top people' does give conservation an air of respectability that makes it acceptable to the Establishment, which without it might look upon the movement as merely the hobby-horse of cranks.

The threat of extermination to endangered species is serious enough in reality without attempts to rouse peoples' emotions by exaggerating. As mentioned in Chapter 6, it seems less than fair dealing to extend the lists of threatened or extinct species by including numerous subspecies or local races, some of which are not taxonomically valid; a hand overplayed may well lose the trick.

This is not the place to go into a detailed history of conservation, but the devoted work of those who founded the International Union for the Conservation of Nature and Natural Resources must be mentioned. This body, with headquarters in Switzerland, has made the public aware of the dangers to the continued existence of much wildlife, and has convinced some part of public opinion of the desirability of trying to save some of it before it is too late. It has, moreover, sponsored scientific investigation into the problems of conservation. This important activity will continue to produce the knowledge without which sound management of wildlife is impossible; it encourages the hope that much that seemed to be travelling fast along the road to destruction may be preserved.

We must not forget, however, that in Great Britain we have had an official Nature Conservancy for a quarter of a century, and a Ministry of Agriculture and Fisheries and its predecessors for over a century. The latter has for years been carrying out scientific research on the wildlife of the seas, in the interests of the commercial fisheries, to conserve the stocks for maximum sustainable yield. All the maritime European countries, the Americas, Australia and New Zealand, and many African and Asian countries have similar official organisations. Their work basically is to gather scientific knowledge for increasing the food supply without depleting the sources. There have been some successes but many failures, owing to the difficulty of regulating the exploitation of international waters which are traditionally free for all.

The conservation of land animals, or some of them, has also been carried on for many years in countries that have big game animals among their fauna. The game departments of Asian, African and American countries at first acted more as gamekeepers than as conservation agents, destroying the species they regarded as vermin in order to present plenty of desirable targets for the guns and rifles of the sportsmen. Such policies, in addition to being acceptable to public opinion, were almost necessary because much of the money needed for administration of the game departments was derived from the licences sold to game-hunters. As game departments evolved, some areas were set aside as sanctuaries where killing animals was forbidden, so that stocks in the hunting reserves might be replenished. These sanctuaries gave rise in time to the establishment of national parks, now maintained by many countries, and first started in the United States with the dedication to the public of Yellowstone Park in 1872.

The parks have been almost too successful, because the beauty of their scenery and the fascination of their wildlife attract such large numbers of visitors in the age of the automobile that it is becoming more and more necessary to limit the numbers of those who may enjoy them at any one time. They are thus bound to lose some of the freedom of their character as wilderness, because roads, hotels, view-points, public lavatories, and all the amenities demanded by people who want to be in the wild places without being of them, have to be provided – people who in ignorance unwittingly degrade what they have come to see. We return to the recurring theme: there are too many people.

Apart from the larger mammals the bird life of certain countries has for long had some amount of protection. The bird protection legislation of Great Britain has become increasingly strict, until now all birds, with a few special exceptions, are completely protected at all times. Similar laws give part or complete protection to the birds in many other countries, so that the bird fauna has undoubtedly benefited until recent years. Unfortunately there are anomalies that frustrate the good intentions of some countries: a notorious example is the netting as delicacies for the table of great numbers of small birds, from thrushes to tomtits, when they are migrating through some Latin countries on their way to or from the northern lands where they are completely protected. This highlights the problem of the international conservation of birds; the protection of resident species by local law works well, but the mobile migrants change their nationality every time they cross a frontier.

In recent years a new hazard, in addition to the risk of being killed for food or sport, threatens the migrant birds, especially the water birds or wildfowl. The marshes and swamps to which they retire for the winter are more and more being drained and brought into agricultural or industrial use; although the breeding grounds in the arctic may yet be largely undisturbed, their winter quarters are being gradually reduced, so that soon they may have nowhere to go. The wildfowl will be crowded out.

The splendid nature reserve in the Coto Donaña in Spain was recently established to preserve the Guadalquivir marshes as a wintering ground for migrant birds from the north, as well as its interesting resident fauna. The events of 1973 show how difficult the maintenance of reserves can be, and how the best-laid plans may be frustrated. In the late summer great numbers of marsh birds were found dead and dying, no one knew why. It was greatly feared that the winter migrants would also be killed in thousands as soon as they should arrive. At first the neighbouring farmers were thought to have polluted the area by excessive use of insecticides to protect their crops, but investigation was unable to incriminate them. Then the cause was found: a natural outbreak of a virulent strain of bacterial disease had struck the birds, and nothing could be done but wait for it to die out. Man is not always the only threat to wildlife.

Our examination of the way man stands in relation to wildlife, and of what his dealings with it have been in the past, shows that with the increasing complexity of man's social and economic growth the relationship also becomes more complex. No longer can wildlife be conserved merely by refraining from exploiting it; man must now decide whether he prefers to conserve or to let the more conspicuous parts of it – the part that is most highly valued by conservationists – disappear for ever in the interests of expanding populations and economic growth. It is no wonder that the conservationists are in a difficult position because, unless and until such fundamental decisions can be agreed and put into practice internationally, their best endeavours can be no more than palliative, and can give only temporary and local protection to the species they think are in greatest need.

It seems inevitable, however, that wildlife as it existed up to the first half of the twentieth century will for the greater part disappear during the next few decades. Under the best of conditions, as the wilderness is progressively brought under control for human occupation, wildlife will be confined to restricted areas such as national parks. Though it may continue to live in them it will be decreasingly wild under the necessary management and human interference that goes into the maintenance of such parks. The only thing that can preserve wildlife, or at least that minority of conspicuous species commonly meant by the term, is for man to stop his population explosion, and to cease chasing the chimera of an ever-expanding economy.

According to the prophets of doom man is fast racing towards the point where he will destroy not only his environment and its wildlife, but his civilisation and most, perhaps all, of his population. If in the course of his evolution man does destroy himself in wars of atomic weapons, or by other means, or leaves only a remnant turned back into the early stone age, the greater part of wildlife in the widest sense, especially the invertebrates, will go merrily on and not even notice that he has gone. Man began as part of wildlife and, if the doomsters' prophesies come true, in spite of all the wrong he has done it, wildlife will survive long after he can no longer despoil it.

Apart from such fantasies there seems little cause to think that reason will prevail so that man will preserve wildlife as well as himself. Man is capable of reasoning, but men, it seems, are not. Whenever a section of the population thinks its personal interests are threatened it discards any morals it may have, and disregards international agreements as well as national policies for the protection of the environment and the conservation of wildlife, in an attempt to satisfy immediate needs. Quarrels and unrest, both within and between nations, are universal and never-ending; wars go on with ever more destructive weapons, and science devises ever more brutal and subtle ways for men to torture the bodies and minds of their fellows. What hope can there be for wildlife? Man's inhumanity to man is matched only by his brutality to the brutes.

FURTHER READING

Chapter 1.

Asimov, I. 1962. *The Wellsprings of Life.* New York: Signet Science Library.
Bodenheimer, F.S. 1965. *Animal Ecology to Day.* 3ed. The Hague: Junk.
Carrington, R. 1956. *A Guide to Earth History.* London: Chatto & Windus.
Ecologist, Editors of The. 1972. *A Blueprint for Survival.* London: Penguin Books.
Elton, C. 1947. *Animal Ecology.* 3ed. London: Sidgwick & Jackson.
Rensch, B. 1972. *Homo Sapiens: from Man to Demigod.* London: Methuen.
Romer, A.S. 1959. *The Vertebrate Story.* Chicago: University Press.
Strahler, A.N. & Strahler, A.H. 1974. *Introduction to Environmental Science.* New York: Hamilton Publishing Co.
Vita-Finzi, C. 1974. *Recent Earth History.* London: Macmillan.

Chapter 2.

Bates, M. 1961. *The Forest and the Sea.* London: Museum Press.
Bold, H.C. 1964. *The Plant Kingdom.* Englewood Cliffs, N.J.: Prentice-Hall.
Carrington, R. 1963. *A Million Years of Man.* London: Weidenfeld & Nicolson.
Clark, W.E. LeGros, 1959. *The Antecedents of Man.* Edinburgh: University Press.
Coon, C.S. 1965. *The Living Faces of Man.* New York: Knopf.
Corner, E.J.H. 1964. *The Life of Plants.* London: Weidenfeld & Nicolson.
Heiser, C.B., jun. 1973. *Seed to Civilization: The Story of Man's Food.* San Francisco & Reading: W.H. Freeman.
Hvass, E. 1973. *Plants that Feed and Serve Us.* 2ed. London: Blandford Press.
Romer, A.S. 1968. *The Procession of Life.* London: Weidenfeld & Nicolson.
Thomas, W. L., Ed. 1962. *Man's Role in Changing the Face of the Earth.* Chicago: University Press.

Chapter 3.

Cushing, D.H. 1968. *Fisheries Biology, A Study in Population Dynamics.* Madison: University of Wisconsin Press.
Davis, F.M. 1959. *Account of the Fishing Gear of England and Wales.* Fisheries Investigations. Series 2, Vol. 21, No. 8. 4ed. London: HMSO.
Hardy, A.C. 1956. *The Open Sea.* London: Collins.
" 1958. *The Open Sea – Fish and Fisheries.* London: Collins.

Marshall, N.B. 1966. *The Life of Fishes.* London: Weidenfeld & Nicolson.
Packard, A. 1972. *Cephalopods and Fish: the Limits of Convergence.* Biol. Rev. 47, 241.
Radcliffe, W. 1926. *Fishing from the Earliest Times.* 2ed. London: Murray.
Woods, J.D. & Lythgoe, J.N. 1972. *Underwater Science.* London: Oxford University Press.

Chapter 4.

Carrington, R. 1971. *The Mediterranean.* London: Weidenfeld & Nicolson.
King, J.E. 1964. *Seals of the World.* London: British Museum (Natural History).
Mackintosh, N.A. 1965. *The Stocks of Whales.* London: Fishing News (Books).
Matthews, L. Harrison. 1952. *Sea Elephant: the Life and Death of the Elephant Seal.* London: MacGibbon & Kee.
1968. *The Whale.* London: Allen & Unwin.
Murphy, R.C. 1947. *Logbook for Grace.* New York: Macmillan.
Priestly, R., Adie, R. J. & Robin, G de Q. 1964. *Antarctic Research.* London Butterworth.
Scheffer, V.B. 1958. *Seals, Sea Lions and Walruses.* Stanford: University Press; London: Oxford University Press.

Chapter 5.

Armstrong, E. A. 1957. *Bird Display and Behaviour.* 2ed. London: Lindsay Drummond.
Austin, O.L. 1961. *Birds of the World.* London: Paul Hamlyn.
Dorst, J. 1961. *Migrations of Birds.* London: Heinemann.
International Council for Bird Preservation. 1963. Bulletin IX. London.
Lister M. *Bird Watcher's Reference Book.* London: Phoenix House.
Matthews, G.T.V. 1973. *Orientation and Position finding by Birds.* London: Oxford University Press.
Murphy, R.C. 1936. *The Oceanic Birds of South America.* 2 vols. New York: American Museum of Natural History.
Selous, E. 1901. *Bird Watching.* London: Dent.

Chapter 6.

Allen, G.M. 1939. *Bats.* Cambridge, Mass: Harvard University Press.
Dasmann, R.F. 1964. *African Game Ranching.* Oxford: Pergamon.
Elton, C. 1942. *Voles, Mice and Lemmings.* Oxford: Clarendon Press.
Kirmiz, J.P. 1962. *Adaptation to Desert Environment.* London: Butterworth.
Matthews, L. Harrison. 1970,1971. *The life of Mammals.* 2 vols. London: Weidenfeld & Nicolson.

Mayr, E. 1963. *Animal Species and Evolution*. Cambridge, Mass: Harvard University Press.
Schmidt-Neilson, K. 1964. *Desert Animals.* Oxford: University Press.
Zeuner, F.E. 1963. *A History of Domesticated Animals.* London: Hutchinson.

Chapter 7.

Brues, C.T. 1946. *Insect Dietry.* Cambridge, Mass: Harvard University Press.
Burr, M. 1954. *The Insect Legion.* London: James Nisbet.
Burton, J. 1968. *The Oxford Book of Insects.* London: Oxford University Press.
Linsenmaier, W. 1972. *Insects of the World.* New York & London: McGraw Hill.
Skaife, S.H. 1953. *Africam Insect Life.* London: Longman Green.
Snow, K.R. 1974. *Insects and Diseases.* London: Routledge.
Wigglesworth, V.B. 1964. *The Life of Insects.* London: Weidenfeld & Nicolson.

Chapter 8.

Cheng, T.C. 1964. *The Biology of Animal Parasites.* Philadelphia: Saunders.
Cloudsley-Thompson, J.L. 1958. *Spiders, Scorpions, Centipedes and Mites.* London & New York: Pergamon.
Kimberlin, R. 1973. *On the Track of the Slow Viruses.* New Scientist, 57, 600.
Manson-Bahr, P. 1960. *Manson's Tropical Diseases*. 15 ed. London: Cassell.
Noble, E.R. & Noble, G.A. 1961. *Parasitology. The Biology of Animal Parasites.* London: Kimpton.
Smith, E., Chapman, G., Clark, R.B., Nicols, D. & Carthy, J.D. 1971. *The Invertebrate Panorama.* London: Weidenfeld & Nicolson.
Wallace, H.R. 1973. *Nematode Ecology and Plant Disease.* London: Arnold.
Zinsser, H. 1934. *Rats, Lice and History.* Boston: Little, Brown.

Chapter 9.

Dorst, J. 1970. *Before Nature Dies.* London: Collins.
Frith, H.J. 1973. *Wildlife Conservation.* Sydney: Angus & Robertson.
Gregory, R. 1971. *The Price of Amenity: Five Studies in Conservation and Government.* London: Macmillan.
Jaubert, A. & Levy-Leblond, J–M., (Eds), 1973. *Autocritique de la Science.* Paris: Editions du Seuil.
Marshall, A.J. 1966. *The Great Extermination.* London: Heinemann.
Matthews, W.H., Smith, F.E. & Goldberg, E.D. 1971. *Terrestrial and Oceanic Ecosystems.* Cambridge, Mass: M.I.T. Press.
Mellanby, K. 1967. *Pesticides and Pollution.* London: Collins.
Smith, A. 1971. *Mato Grosso.* London: Michael Joseph.

Index

www.ingramcontent.com/pod-product-compliance
Ingram Content Group UK Ltd.
Pitfield, Milton Keynes, MK11 3LW, UK
UKHW021507240926
475702UK00035B/650
9780367416683